Proceedings of the Institute of General Physics
Academy of the Sciences of the USSR
Series Editor: A.M. Prokhorov

Proceedings of the Institute of General Physics
Academy of the Sciences of the USSR

Series Editor: A.M. Prokhorov

Volume 1 **Oceanic Remote Sensing,** Edited by F.V. Bunkin and K.I. Volyak.

Volume 2 **Laser Raman Spectroscopy in Crystals and Gases,** Edited by P.P. Pashinin.

Volume 3 **The Magnetic and Electron Structures of Transition Metals and Alloys,** Edited by V.G. Veselago and L.I. Vinokurova.

Volume 4 **Laser Methods of Defect Investigations in Semiconductors and Dielectrics,** Edited by A.A. Manenkov.

Volume 5 **Fiber Optics,** Edited by E.M. Dianov.

Volume 6 **The Nonlinear Optics and Acoustics of Fluids,** Edited by F.V. Bunkin.

Volume 7 **Formation and Control of Optical Wavefronts,** Edited by P.P. Pashinin.

Volume 8 **Problems of Lithography in Microelectronics,** Edited by T.M. Makhviladze.

Volume 9 **Selective Laser Spectroscopy of Activated Crystals and Glasses,** Edited by V.V. Osiko.

Proceedings of the Institute of General Physics
Academy of the Sciences of the USSR
Series Editor: A.M. Prokhorov
Volume 5

FIBER OPTICS

Edited by Ye. M. Dianov

Translated by Kevin S. Hendzel

NOVA SCIENCE PUBLISHERS
COMMACK

NOVA SCIENCE PUBLISHERS
283 Commack Road
Suite 300
Commack, New York 11725

This book is being published under exclusive English Language rights granted to Nova Science Publishers, Inc. by the All-Union Copyright Agency of the USSR (VAAP).

Library of Congress Cataloging-in-Publication Data

Volokonnaîa optika. English.
Fiber optics.

(Proceedings of the Institute of General Physics of the Academy of Sciences of the USSR ; v. 5)
Translation of: Volokonnaîa optika.
Bibliography: p.
Includes index.
1. Fiber optics. I. Dianov, E. M. II. Series: Trudy Instituta obshcheĭ fiziki. English ; v. 5.
TA1800.V6413 1988 621.36'92 88-6623
ISBN 0-941743-16-0

The original Russian-language version of this book was published by Nauka Publishing House in 1987.

Printed in the United States of America

CONTENTS

MECHANICAL STRENGTH OF LONG-LENGTH VITREOUS SILICA GLASS OPTICAL FIBERS 79

V.A. Bogatyrev, M.M. Bubnov, N.N. Vechkanov
A.N. Gur'yanov, S.L. Semenov

TEMPERATURE RESISTANT FIBER-OPTIC MODULES 97

A.A. Abramov, M.M. Bubnov, N.N. Vechkanov, A.N. Gur'yanov,
A.S. Konov, V.N. Myakov, B.B. Troitskiy, A.G. Shchebunyaev

OPTICAL PROPERTIES OF GERMANIUM SILICATE GLASS FOR LOW-LOSS OPTICAL FIBERS 111

V.M. Mashinskiy

DATA TRANSMISSION BY SPECTRAL- [WAVELENGTH-] DIVISION MULTIPLEXING 191
A. A. Kuznetsov

ULTRASHORT PULSES IN FIBER STIMULATED LIGHT SCATTERING LASERS 203
Ye.M. Dianov, A.N. Pilipetskiy, A.M. Prokhorov, V.N. Serkin

SUBJECT INDEX 216

The Dispersion Characteristics of Three-Layer Elliptical Optical Fibers

A.S. Belanov, Ye.M. Dianov, V.I. Krivenkov

Abstract: A method is proposed for calculating the dispersion characteristics of three-layer elliptical optical fibers which is then used to determine the retardation factors of the phase and group velocities as well as the waveguide dispersion of the two lower order orthogonally-polarized modes. The optical fiber parameters are established for the case where the difference of the retardation factors of the phase velocities of such modes is maximized with equal group velocities.

Introduction

A number of optical fiber applications such as fiber-optic interferometers, coherent communications links, etc. require conservation of the mode polarization over significant optical fiber lengths. Axiosymmetric refractive index optical fibers of circular cross-section cannot conserve the direction of mode polarization over relatively long optical fiber lengths, since unavoidable irregularities convert the working mode to an orthogonally-polarized mode, since the phase constants of these modes are identical. In order to conserve polarization it is possible to use elliptical waveguides in which the degeneration of the orthogonally-polarized modes is eliminated. The phase constants of these modes are different and, consequently, their conversion by irregularities is reduced.

We know that there are stringent requirements to conserve the linearity of modal polarization. In single-mode optical fibers of circular cross-section in single polarization operation the mode field always possesses undesirable orthogonally-polarized components. Study

[1] has investigated the polarization linearity of the He_{11} mode in dielectric waveguides. We carried out an analogous investigation for dual- and three-layer optical fibers. We may characterize the non-linearity of mode polarization by means of the depolarization coefficient

$$k_d = \sum_{i=1}^{n} |\Delta P_i| / P_\Sigma , \tag{1}$$

where P_Σ is the total mode power conveyed in n optical fiber layers; ΔP_i is the difference between the power that would be generated by the linearly-polarized components of the mode field and the actual mode power in the i optical fiber layer.

Our investigation for dual-layer step refractive index optical fibers made it possible to establish that maximum depolarization is observed when $V = 2\pi a \lambda^{-1} (n_1^2 - n_2^2)^{1/2} = 1{,}872,$ when

$$k_{d\,\text{макс}} = 8{,}072 \cdot 10^{-3} (\epsilon_1 - 1)^{1{,}982}, \tag{2}$$

or, by expressing $k_{d\ \text{макс}}$ in decibels,

$$k_{d\,\text{макс}} = -20{,}93 + 19{,}82 \lg(\epsilon_1 - 1). \tag{3}$$

When $V = 2.4$ the depolarization coefficient is

$$k_d = 7{,}16 \cdot 10^{-3} (\epsilon_1 - 1)^{1{,}978}, \tag{4}$$

or, expressed in decibels

$$k_d = -21{,}45 + 19{,}78 \lg(\epsilon_1 - 1), \tag{5}$$

where $\varepsilon_1 = n_1^2/n_2^2$. When $n_1 - n_2 \ll 1$ we may take $\varepsilon_1 - 1 = 2\Delta$, where $\Delta = (n_1 - n_2)/n_2$ is the relative differential of the refractive indices of the optical fiber core and cladding.

The derived formulae for k_d are valid for $\varepsilon_1 < 1.1$ with an error of less than 1% and when $1.1 < \varepsilon_1 < 2$ with an error of less than 5%. Consequently in single-mode optical fibers with $\Delta = 0.2$-0.8% it is possible to achieve a maximum depolarization coefficient of $k_d = (69$-$63)$ dB.

A number of optical fiber applications require high mode polarization linearity. The most appropriate optical fibers in these cases are three-layer elliptical optical fibers.

The present study investigates the dispersion characteristics of the three-layer elliptical optical fibers proposed previously by the present authors. Compared to the dual-layer elliptical optical fibers, three-layer elliptical optical fibers have the same advantages that three-layer optical fibers of circular cross-section [2-5] have over dual-layer fibers, specifically: significant core cross-sections in single-mode operation; improved dispersion characteristics that compensate the influence of both positive and negative material dispersion due to the waveguide dispersion of opposite sign; the capability for working mode selection by increasing losses for the nearest higher-order modes; and the capability to transmit powerful signals without significant nonlinear distortion in circular three-layer elliptical optical fibers.

It is significant that *W*-type three layer elliptical optical fibers may in principle allow single-polarization, single-mode operation when the lower order mode with the undesirable orthogonal polarization cannot travel through the optical fiber. It is possible to create conditions in the three-layer elliptical optical fibers in which modes of both polarizations may travel where the mode with the undesirable polarization may acquire elevated dielectric losses or emission losses, which, in addition to the differential in their phase constants, will help to reduce the coupling between these modes caused by the irregularities and will conserve working mode polarization [6].

Solving the problem of wave propagation in a dielectrical elliptical waveguide is more complex than the analogous problem for a dielectric waveguide of circular cross-section. Therefore the initial studies devoted to the analysis of an elliptical dielectric waveguide appeared only in the early 1960's [7-9] at the same time that the first analysis of circular dielectric waveguides was carried out in 1910 [10].

The first study devoted to a theoretical investigation of dual-layer elliptical optical fibers in the small difference approximation of the indices of the core and the cladding, Yeh [11] determined the normalized propagation constants of three even and odd modes. Later Rengarajan and Lewis [12] investigated three-layer elliptical optical fibers.

One feature of the solution of the problem of wave propagation in elliptical optical fibers was that Mathieu's function was used to represent the longitudinal field components. The use of such components in the analysis of three-layer elliptical optical fibers has made it possible to successfully analyze only optical fibers in which the ellipses of the layer boundaries are confocal [12].

The solution to the more general problem of wave propagation in three-layer elliptical waveguides in which the ellipses of the interfaces are arbitrary and not necessarily confocal is of more practical interest. For such waveguides it is best to use cylindrical functions to represent the longitudinal field components.

1. Formulation of the Problem

The present study proposes a method of calculating the dispersion characteristics of a three-layer elliptical waveguide (Fig. 1). The waveguide core is an elliptical cylinder with a refractive index

n_1; a_1, b_1 are the semi-major and semi-minor axes of the ellipse of the interface between the core and the intermediate cladding whose refractive index is equal to n_2; a_2, b_2 are the semi-major and semi-minor axes of the ellipse of the interface of the intermediate cladding and the outer cladding. The outer cladding whose refractive index is n_3 is assumed to be sufficiently thick so that the influence of the external environment on the mode field may be neglected. The present study employs the following conventions: $\epsilon_1 = n_1^2/n_2^2$, $\epsilon_3 = n_3^2/n_2^2$, $c = b_1/b_2$. The eccentricities of the ellipses of the interfaces between the layers are determined by the relations $e_1 = (1 - b_1^2/a_1^2)^{1/2}$, $e_2 = (1 - b_2^2/a_2^2)^{1/2}$.

In an electromagnetic wave (mode) traveling on the optical fiber axis (z) all field components have a dependence on the time t and longitudinal coordinate z as $\exp[i(\omega t-\gamma z)]$, where γ is the longitudinal propagation constant of the mode, while ω is its frequency. Specifically we may represent the longitudinal electric field component as $E_z \exp[i(\omega t - \gamma z)]$, and the magnetic field longitudinal component as $H_z \exp[i(\omega t - \gamma z)]$, where E_z, H_z are the transverse coordinate functions.

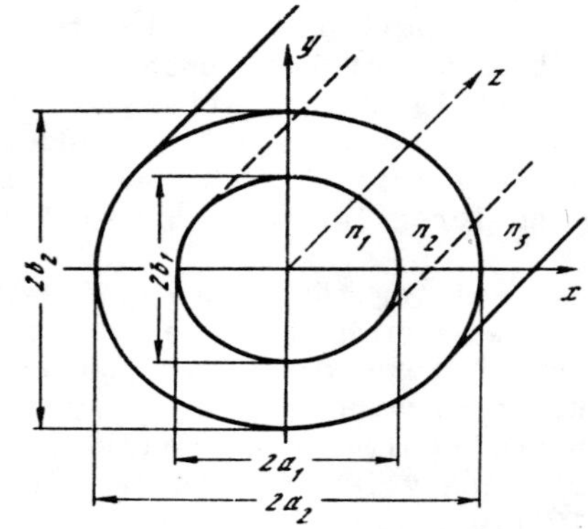

Fig. 1. Three-layer elliptical optical fiber

It follows from Maxwell's equations that E_z, H_z satisfy the two-dimensional wave equation

$$(\nabla^2 + k^2) \begin{Bmatrix} E_z \\ H_z \end{Bmatrix} = 0, \tag{6}$$

where ∇^2 is a two-dimensional Laplacian operator that in the polar coordinate system (r, φ) takes the form

$$\nabla^2 = \frac{1}{r}\frac{\partial}{\partial r}\left(r\frac{\partial}{\partial r}\right) + \frac{1}{r^2}\frac{\partial^2}{\partial \varphi^2}; \tag{7}$$

$k = (n^2 k_0^2 - \gamma^2)^{1/2}$ – is the transverse wave number; $k_0 = \omega/c$ is the wave number in a vacuum; n is the refractive index of the medium.

We remember that the polar coordinates (r, φ) are related to the Cartesian coordinates (x, y) by the relations

$$x = r\cos\varphi,\ y = r\sin\varphi. \tag{8}$$

In the optical fiber core $n = n_1$ and

$$k_1^2 = n_1^2 k_0^2 - \gamma^2, \tag{9}$$

and in the intermediate cladding $n = n_2$ and

$$k_2^2 = n_2^2 k_0^2 - \gamma^2, \tag{10}$$

in the outer cladding $n = n_3$ and

$$k_3^2 = n_3^2 k_0^2 - \gamma^2 . \tag{11}$$

We will introduce the dimensionless transverse wave numbers

$$x_1 = b_2 | k_1 |, x_2 = b_2 |k_2|, x_3 = b_2 |k_3| \tag{12}$$

and the dimensionless characteristic optical fiber parameter

$$V = b_1 k_0 |n_1^2 - n_2^2|^{1/2}. \tag{13}$$

We will characterize the phase velocity of the mode by means of the phase retardation coefficient

$$U = \gamma / k_0 n_2 . \tag{14}$$

Using the previous conventions as well as (13) and (14) we obtain from (9)-(11):

$$\begin{aligned} b_2^2 k_1^2 &= V^2 (\epsilon_1 - U^2)(c^2 |\epsilon_1 - 1|)^{-1}, \\ b_2^2 k_2^2 &= V^2 (1 - U^2)(c^2 |\epsilon_1 - 1|)^{-1}, \\ b_2^2 k_3^2 &= V^2 (\epsilon_3 - U^2)(c^2 |\epsilon_1 - 1|)^{-1}. \end{aligned} \tag{15}$$

The problem becomes finding the dependence of the phase velocity retardation coefficient U on the characteristic parameter of the optical fiber V.

We may obtain other dispersion characteristics of the optical fiber by operating on the $U(V)$ function with the appropriate linear differential operator. For example, the group velocity retardation coefficient U_{rp} is determined by the formula

$$U_{rp} = U + V dU/dV. \tag{16}$$

2. Longitudinal Mode Components

Since E_z and H_z are continuous and periodic (with a period of 2π) functions of φ, they may be expanded into a Fourier series, i.e., represented as

$$\begin{Bmatrix} E_z \\ H_z \end{Bmatrix} = \sum_{m=0}^{\infty} \begin{Bmatrix} A_E \\ A_H \end{Bmatrix} \cos m\varphi + \sum_{m=1}^{\infty} \begin{Bmatrix} B_E \\ B_H \end{Bmatrix} \sin m\varphi,$$

where A_E, A_H, B_E, B_H are functions of r.

It follows from the symmetry properties of the optical fiber, specifically its invariance with respect to rotation by π and specular reflection in the plane passing through the optical fiber axis and one of the elliptical axes of the interfaces between layers that E_z and H_z are invariant with respect to the transformation $\varphi \to \varphi + \pi$ and if the coordinate system is selected so that the origin φ coincides with one of the axes of the ellipses of the interfaces between the layers, then E_z, H_z will be invariant with respect to the transformation $\varphi = -\varphi$. Hence in this coordinate system E_z, H_z may be represented in one of the following four series:

$$\sum_{m=0}^{\infty} A_m \cos(2m+1)\varphi, \quad \sum_{m=0}^{\infty} A_m \sin(2m+1)\varphi, \quad \sum_{m=0}^{\infty} A_m \cos 2m\varphi, \quad \sum_{m=1}^{\infty} A_m \sin 2m\varphi,$$

where A_m are functions of r.

Since the transverse field components will also be invariant with respect to these transformations, then if E_z is represented as a series in terms of sin, H_z will be represented as a series in terms of cos and vice versa.

Thus, if $H_z = \sum_{m=0}^{\infty} B_m \cos(2m+l)\varphi$, then $E_z = \sum_{m=1}^{\infty} A_m \sin(2m-l)\varphi$, which corresponds to even $HE_{2m-l,n}$ and $EH_{2m-l,n}$ modes (l of 0 or 1; $m = 1, 2, \ldots$; $n = 1, 2, \ldots$).

The odd $HE_{2m-l,n}$ and $EH_{2m-l,n}$ modes are determined by representing E_z and H_z as

$$E_z = \sum_{m=0}^{\infty} A_m \cos(2m+l)\varphi, \quad H_z = \sum_{m=1}^{\infty} B_m \sin(2m-l)\varphi.$$

On this basis and also accounting for the fact that the longitudinal field components will satisfy wave equation (6) we will represent E_z and H_z in a optical fiber for even modes in the following form:

in the core

$$E_{z1}(r,\varphi) = \begin{cases} \sum_{m=1}^{\infty} A_m^1 J_{2m-l}(rk_1)\sin(2m-l)\varphi, & k_1^2 \geqslant 0, \\ \sum_{m=1}^{\infty} A_m^1 I_{2m-l}(r|k_1|)\sin(2m-l)\varphi, & k_1^2 < 0, \end{cases}$$

$$H_{z1}(r,\varphi) = \begin{cases} \sum_{m=0}^{\infty} B_{m+1}^1 J_{2m+l}(rk_1)\cos(2m+l)\varphi, & k_1^2 \geqslant 0, \\ \sum_{m=0}^{\infty} B_{m+1}^1 I_{2m+l}(r|k_1|)\cos(2m+l)\varphi, & k_1^2 < 0; \end{cases} \tag{17}$$

in the intermediate cladding

$$E_{z2}(r,\varphi)=\begin{cases}\sum\limits_{m=1}^{\infty}[A_m^2 J_{2m-l}(rk_2)+\widetilde{A}_m^2 N_{2m-l}(rk_2)]\sin(2m-l)\varphi, & k_2^2\geqslant 0,\\ \sum\limits_{m=1}^{\infty}[A_m^2 I_{2m-l}(r|k_2|)+\widetilde{A}_m^2 K_{2m-l}(r|k_2|)]\sin(2m-l)\varphi, & k_2^2<0,\end{cases}$$

$$H_{z2}(r,\varphi)=\begin{cases}\sum\limits_{m=0}^{\infty}[B_{m+1}^2 J_{2m+l}(rk_2)+\widetilde{B}_{m+1}^2 N_{2m+l}(rk_2)]\cos(2m+l)\varphi, & k_2^2\geqslant 0,\\ \sum\limits_{m=0}^{\infty}[B_{m+1}^2 I_{2m}(r|k_2|)+\widetilde{B}_{m+1}^2 K_{2m+l}(r|k_2|)]\cos(2m+l)\varphi, & k_2^2<0;\end{cases} \tag{18}$$

in the outer cladding

$$E_{z3}(r,\varphi)=\sum_{m=1}^{\infty}A_m^3 K_{2m-l}(r|k_3|)\sin(2m-l)\varphi,$$

$$H_{z3}(r,\varphi)=\sum_{m=0}^{\infty}B_{m+1}^3 K_{2m+l}(r|k_3|)\cos(2m+l)\varphi, \tag{19}$$

where l is either 0 or 1 depending on which even modes are considered, while J_m, I_m, N_m, K_m are cylindrical functions.

This representation accounts for the fact that E_z and H_z will be finite when $r = 0$ while when $r \to \infty$ they will decay exponentially, for which we require $k_3^2 < 0$. If we carry out the substitution of E_{zi} by H_{zi} and H_{zi} by E_{zi} ($i = 1, 2, 3$) in (17)-(19), we obtain the representation for E_z and H_z corresponding to the odd modes.

3. Dispersion Equation

We will first introduce certain relations necessary for further transformations.

Each ellipse in the plane may be unambiguously assigned to the (ξ, η) elliptical coordinate system so that a certain coordinate line (ξ_0, η) will be a given ellipse. Assume we have an ellipse whose semi-major and semi-minor axes are equal to a and b. Fig. 2 shows the elliptical coordinate system (ξ, η) for which the given ellipse is a certain coordinate line (ξ_0, η), where $\xi_0 = \mathrm{arth}(b/a)$.

The elliptical coordinates are related to the Cartesian coordinates by the following relations:

$$x=\sigma\,\mathrm{ch}\,\xi\cos\eta,\quad y=\sigma\,\mathrm{sh}\,\xi\sin\eta, \tag{20}$$

where $\sigma = (a^2 - b^2)^{1/2}$.

It is easily demonstrated that it follows from relations (8), (20) that

$$r=\sigma(\mathrm{sh}^2\xi+\cos^2\eta)^{1/2},\quad \varphi=\mathrm{arctg}(\mathrm{th}\,\xi\,\mathrm{tg}\,\eta). \tag{21}$$

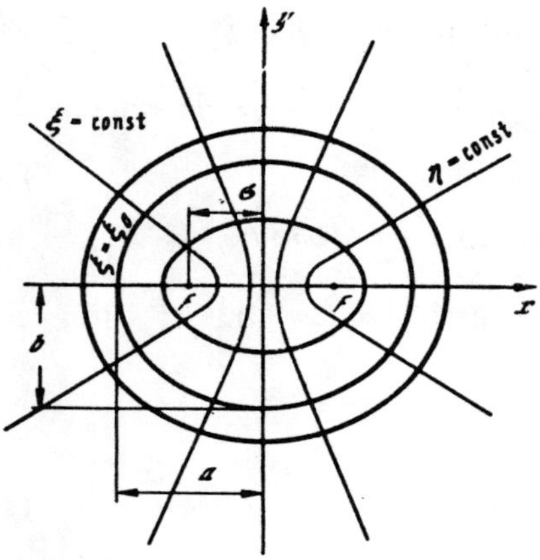

Fig. 2. Elliptical cylindrical coordinate system.

We obtain from the relations

$$\frac{\partial}{\partial\xi} = \frac{\partial\varphi}{\partial\xi}\frac{\partial}{\partial\varphi} + \frac{\partial r}{\partial\xi}\frac{\partial}{\partial r}, \quad \frac{\partial}{\partial\eta} = \frac{\partial\varphi}{\partial\eta}\frac{\partial}{\partial\varphi} + \frac{\partial r}{\partial\eta}\frac{\partial}{\partial r}$$

and relations (21)

$$\frac{\partial}{\partial\xi} = \frac{\sin 2\varphi}{\mathrm{sh}2\xi}\frac{\partial}{\partial\varphi} + \frac{\sigma^2}{2r}\,\mathrm{sh}2\xi\frac{\partial}{\partial r},$$

$$\frac{\partial}{\partial\eta} = \frac{\sigma^2}{2r^2}\mathrm{sh}2\xi\frac{\partial}{\partial\varphi} - \frac{r\sin 2\varphi}{\mathrm{sh}2\xi}\frac{\partial}{\partial r}.$$

For the given ellipse (see Fig. 2) we have

$$\left.\frac{\partial}{\partial\xi}\right|_{\xi=\xi_0} = \frac{1}{\sqrt{1-e_0^2}}\left.\left(\frac{e_0^2}{2}\sin 2\varphi\frac{\partial}{\partial\varphi} + \frac{b^2}{\rho_0}\frac{\partial}{\partial r}\right)\right|_{r=\rho_0},$$
$$\left.\frac{\partial}{\partial\eta}\right|_{\xi=\xi_0} = \frac{1}{\sqrt{1-e_0^2}}\left.\left(\frac{b^2}{\rho_0^2}\frac{\partial}{\partial\varphi} - \frac{e_0^2\rho_0}{2}\sin 2\varphi\frac{\partial}{\partial r}\right)\right|_{r=\rho_0}, \tag{22}$$

where $e_0 = (1 - b^2/a^2)^{1/2}$ is the eccentricity of the given ellipse, while $\rho_0 = b(1 - e_2^2\cos^2\varphi)^{-1/2}$.

We introduce the functions

$$S_m(e,\varphi) = \sin m\varphi - E\sin(m-2)\varphi,\quad \tilde{S}_m(e,\varphi) = E\sin(m+2)\varphi - \sin m\varphi,$$
$$C_m(e,\varphi) = \cos m\varphi - E\cos(m-2)\varphi,\quad \tilde{C}_m(e,\varphi) = E\cos(m+2)\varphi - \cos m\varphi, \tag{23}$$

where $e = e^2(2 - e^2)^{-1}$.

We will determine the function $\Phi_m = J_m(rx)\begin{Bmatrix}\sin m\varphi\\ \cos m\varphi\end{Bmatrix}$. After substituting this function into relation (22) we may demonstrate that

$$\left.\frac{\partial\Phi_m}{\partial\xi}\right|_{\xi=\xi_0} = Rx\left[J_{m-1}(\rho_0 x)\begin{Bmatrix}S_m(e_0,\varphi)\\ C_m(e_0,\varphi)\end{Bmatrix} + J_{m+1}(\rho_0 x)\begin{Bmatrix}\tilde{S}_m(e_0,\varphi)\\ \tilde{C}_m(e_0,\varphi)\end{Bmatrix}\right],$$
$$\left.\frac{\partial\Phi_m}{\partial\eta}\right|_{\xi=\xi_0} = Rx\left[J_{m-1}(\rho_0 x)\begin{Bmatrix}C_m(e_0,\varphi)\\ -S_m(e_0,\varphi)\end{Bmatrix} - J_{m+1}(\rho_0 x)\begin{Bmatrix}\tilde{C}_m(e_0,\varphi)\\ -\tilde{S}_m(e_0,\varphi)\end{Bmatrix}\right], \tag{24}$$

where $R = 0{,}5\rho_0(1 - e_0^2/2)(1 - e_0^2)^{-1/2}$. If $\Phi_m = I_m(rx)\begin{Bmatrix}\sin m\varphi\\ \cos m\varphi\end{Bmatrix}$, then

$$\left.\frac{\partial\Phi_m}{\partial\xi}\right|_{\xi=\xi_0} = Rx\left[I_{m-1}(\rho_0 x)\begin{Bmatrix}S_m(e_0,\varphi)\\ C_m(e_0,\varphi)\end{Bmatrix} - I_{m+1}(\rho_0 x)\begin{Bmatrix}\tilde{S}_m(e_0,\varphi)\\ \tilde{C}_m(e_0,\varphi)\end{Bmatrix}\right],$$
$$\left.\frac{\partial\Phi_m}{\partial\eta}\right|_{\xi=\xi_0} = Rx\left[I_{m-1}(\rho_0 x)\begin{Bmatrix}C_m(e_0,\varphi)\\ -S_m(e_0,\varphi)\end{Bmatrix} + I_{m+1}(\rho_0 x)\begin{Bmatrix}\tilde{C}_m(e_0,\varphi)\\ -\tilde{S}_m(e_0,\varphi)\end{Bmatrix}\right]. \tag{25}$$

Certain relations analogous to (24) are valid for the function $N_m(rx)$ $\begin{Bmatrix} \sin m\varphi \\ \cos m\varphi \end{Bmatrix}$ while relations analogous to (25) are valid for the function $K_m(rx) \begin{Bmatrix} \sin m\varphi \\ \cos m\varphi \end{Bmatrix}$ – only with a reversed sign.

It follows from Maxwell's equations in the coordinate system of the elliptical cylinder (ξ, η, z) whose z axis coincides with the optical fiber axis with η read from the major axis of the interface ellipses between the optical fiber layers that

$$E_\eta = \frac{Q_E}{k^2}\left(U\frac{\partial E_z}{\partial \eta} - \frac{\partial H_z}{\partial \xi}\right),\quad H_\eta = \frac{Q_H}{k^2}\left(\epsilon\frac{\partial E_z}{\partial \xi} + U\frac{\partial H_z}{\partial \eta}\right), \tag{26}$$

where

$$k^2 = \begin{cases} k_1^2 & \text{in the optical fiber core,} \\ k_2^2 & \text{in the intermediate cladding} \\ k_3^2 & \text{in the outer cladding} \end{cases} \tag{27}$$

$$\epsilon = \begin{cases} \epsilon_1 & \text{in the optical fiber core} \\ 1 & \text{in the intermediate cladding} \\ \epsilon_3 & \text{in the outer cladding} \end{cases} \tag{28}$$

Q_E and Q_H are certain continuous coordinate functions that cancel upon further manipulation.

Selecting for each interface ellipse between the layers the appropriate coordinate system of the elliptical integral and accounting for relations (26)-(28) as well as the boundary conditions and that $\partial e_z/\partial\eta$ and $\partial H_z/\partial\eta$ are continuous functions (since E_z and H_z are continuous functions), we obtain the equation system

$$\begin{aligned} &E_{z1}[b_2\rho_1(\varphi),\varphi] = E_{z2}[b_2\rho_1(\varphi),\varphi],\quad H_{z1}[b_2\rho_1(\varphi),\varphi] = H_{z2}[b_2\rho_1(\varphi),\varphi], \\ &E_{z3}[b_2\rho_2(\varphi),\varphi] = E_{z2}[b_2\rho_2(\varphi),\varphi],\quad H_{z3}[b_2\rho_2(\varphi),\varphi] = H_{z2}[b_2\rho_2(\varphi),\varphi], \end{aligned} \tag{29}$$

$$\frac{1}{k_1^2}\left(\frac{\partial H_{z1}}{\partial \xi}\right)\bigg|_{\xi=\xi_1} = \frac{1}{k_2^2}\left(\frac{\partial H_{z2}}{\partial \xi}\right)\bigg|_{\xi=\xi_1} - U\left(\frac{1}{k_2^2} - \frac{1}{k_1^2}\right)\left(\frac{\partial E_{z2}}{\partial \eta}\right)\bigg|_{\xi=\xi_1},$$

$$\frac{\epsilon_1}{k_1^2}\left(\frac{\partial E_{z1}}{\partial \xi}\right)\bigg|_{\xi=\xi_1} = \frac{1}{k_2^2}\left(\frac{\partial E_{z2}}{\partial \xi}\right)\bigg|_{\xi=\xi_1} + U\left(\frac{1}{k_2^2} - \frac{1}{k_1^2}\right)\left(\frac{\partial H_{z2}}{\partial \eta}\right)\bigg|_{\xi=\xi_1},$$

$$\frac{1}{k_3^2}\left(\frac{\partial H_{z3}}{\partial \xi}\right)\bigg|_{\xi=\xi_2} = \frac{1}{k_2^2}\left(\frac{\partial H_{z2}}{\partial \xi}\right)\bigg|_{\xi=\xi_2} - U\left(\frac{1}{k_2^2} - \frac{1}{k_3^2}\right)\left(\frac{\partial E_{z2}}{\partial \eta}\right)\bigg|_{\xi=\xi_2},$$

$$\frac{\epsilon_3}{k_3^2}\left(\frac{\partial E_{z3}}{\partial \xi}\right)\bigg|_{\xi=\xi_2} = \frac{1}{k_2^2}\left(\frac{\partial E_{z2}}{\partial \xi}\right)\bigg|_{\xi=\xi_2} + U\left(\frac{1}{k_2^2} - \frac{1}{k_3^2}\right)\left(\frac{\partial H_{z2}}{\partial \eta}\right)\bigg|_{\xi=\xi_2},$$

where

$$\rho_1 = c(1 - e_1^2 \cos^2 \varphi)^{-1/2}; \quad \rho_2 = (1 - e_2^2 \cos^2 \varphi)^{-1/2};$$

$$\xi_1 = \operatorname{arth}(b_1/a_1); \quad \xi_2 = \operatorname{arth}(b_2/a_2); \quad c = b_1/b_2.$$

We will introduce an additional definition. Assume an infinite series of functions $f_1(\varphi)$, $f_2(\varphi)\ldots$, is given such that each function from this sequence may be represented in the form of the series

$$f_j(\varphi) = \sum_{i=1}^{\infty} a_{ij} \cos(2i - 2 + l)\varphi, \quad j = 1, 2, \ldots, \tag{30}$$

or in the form of the series

$$f_j(\varphi) = \sum_{i=1}^{\infty} a_{ij} \sin(2i - l)\varphi, \quad j = 1, 2, \ldots, \tag{31}$$

where l takes either 0 or 1.

We will assign the sequence of functions $f_1(\varphi)$, $f_2(\varphi)$, ... to matrix A with elements a_{ij} and will call this sequence of functions the sequence of functions determining the matrix A.

Accounting for the form of the functions E_z and H_z in the optical fiber (17)-(19) as well as relations (24), (25) we may easily demonstrate that any equation from equation system (29) may be expanded into a Fourier series of the type in (30) or (31) depending on which of the equations is selected.

Representing the unknown coefficients in equation system (29) as the vectors

$$\begin{aligned}
\mathbf{A}_1 &= (A_1^1, A_2^1, \ldots), & \mathbf{B}_1 &= (B_1^1, B_2^1, \ldots), \\
\mathbf{A}_2 &= (A_1^2, A_2^2, \ldots), & \mathbf{B}_2 &= (B_1^2, B_2^2, \ldots), \\
\widetilde{\mathbf{A}}_2 &= (\widetilde{A}_1^2, \widetilde{A}_2^2, \ldots), & \widetilde{\mathbf{B}}_2 &= (\widetilde{B}_1^2, \widetilde{B}_2^2, \ldots), \\
\mathbf{A}_3 &= (A_1^3, A_2^3, \ldots), & \mathbf{B}_3 &= (B_1^3, B_2^3, \ldots)
\end{aligned} \tag{32}$$

and expanding each of the equations of this system into the corresponding Fourier series (either (30) or (31)) we obtain an infinite system of equations that may be represented in this form:

$$\begin{aligned}
&E_1 \mathbf{A}_1 = E_2^1 \mathbf{A}_2 + \widetilde{E}_2^1 \widetilde{\mathbf{A}}_2, \quad H_1 \mathbf{B}_1 = H_2^1 \mathbf{B}_2 + \widetilde{H}_2^1 \widetilde{\mathbf{B}}_2, \\
&E_3 \mathbf{A}_3 = E_2^3 \mathbf{A}_2 + \widetilde{E}_2^3 \widetilde{\mathbf{A}}_2, \quad H_3 \mathbf{B}_3 = H_2^3 \mathbf{B}_2 + \widetilde{H}_2^3 \widetilde{\mathbf{B}}_2, \\
&H_{\xi 1} \mathbf{B}_1 = H_{\xi 2}^1 \mathbf{B}_2 + \widetilde{H}_{\xi 2}^1 \widetilde{\mathbf{B}}_2 - E_{\eta 2}^1 \mathbf{A}_2 - \widetilde{E}_{\eta 2}^1 \widetilde{\mathbf{A}}_2, \\
&E_{\xi 1} \mathbf{A}_1 = E_{\xi 2}^1 \mathbf{A}_2 + \widetilde{E}_{\xi 2}^1 \widetilde{\mathbf{A}}_2 + H_{\eta 2}^1 \mathbf{B}_2 + \widetilde{H}_{\eta 2}^1 \widetilde{\mathbf{B}}_2, \\
&H_{\xi 3} \mathbf{B}_3 = H_{\xi 2}^3 \mathbf{B}_2 + \widetilde{H}_{\xi 2}^3 \widetilde{\mathbf{B}}_2 - E_{\eta 2}^3 \mathbf{A}_2 - \widetilde{E}_{\eta 2}^3 \widetilde{\mathbf{A}}_2, \\
&E_{\xi 3} \mathbf{A}_3 = E_{\xi 2}^3 \mathbf{A}_2 + \widetilde{E}_{\xi 2}^3 \widetilde{\mathbf{A}}_2 + H_{\eta 2}^3 \mathbf{B}_2 + \widetilde{H}_{\eta 2}^3 \widetilde{\mathbf{B}}_2,
\end{aligned} \tag{33}$$

or in a more compact form

$$\begin{aligned}
&E_i \mathbf{A}_i = E_2^i \mathbf{A}_2 + \widetilde{E}_2^i \widetilde{\mathbf{A}}_2, \quad H_i \mathbf{B}_i = H_2^i \mathbf{B}_2 + \widetilde{H}_2^i \widetilde{\mathbf{B}}_2, \\
&H_{\xi i} \mathbf{B}_i = H_{\xi 2}^i \mathbf{B}_2 + \widetilde{H}_{\xi 2}^i \widetilde{\mathbf{B}}_2 - E_{\eta 2}^i \mathbf{A}_2 - \widetilde{E}_{\eta 2}^i \widetilde{\mathbf{A}}_2, \\
&E_{\xi i} \mathbf{A}_i = E_{\xi 2}^i \mathbf{A}_2 + \widetilde{E}_{\xi 2}^i \widetilde{\mathbf{A}}_2 + H_{\eta 2}^i \mathbf{B}_2 + \widetilde{H}_{\eta 2}^i \widetilde{\mathbf{B}}_2,
\end{aligned} \tag{34}$$

where $E_i, H_i, E_2^i, \widetilde{E}_2^i, H_2^i, \widetilde{H}_2^i, E_{\xi i}, H_{\xi i}, E_{\xi 2}^i, \widetilde{E}_{\xi 2}^i, H_{\xi 2}^i, \widetilde{H}_{\xi 2}^i, E_{\eta 2}^i, \widetilde{E}_{\eta 2}^i, H_{\eta 2}^i, \widetilde{H}_{\eta 2}^i$ are matrices, $i = 1, 3$.

Using (12), (17)-(19), (23)-(25), (29), we may write the determining sequences of functions for all matrices entering into equation system (33).

For even modes

$$E_1 \Leftrightarrow \begin{cases} J_{2m-l}\sin(2m-l)\varphi, & k_1^2 \geqslant 0, \\ I_{2m-l}\sin(2m-l)\varphi, & k_1^2 < 0, \end{cases}$$
$$H_1 \Leftrightarrow \begin{cases} J_{2m+l}\cos(2m+l)\varphi, & k_1^2 \geqslant 0, \\ I_{2m+l}\cos(2m+l)\varphi, & k_1^2 < 0, \end{cases} \tag{35}$$

where the argument of the cylindrical functions is $\rho_1 x_1$;

$$E_2^{2i-1} \Leftrightarrow \begin{cases} J_{2m-l}\sin(2m-l)\varphi, & k_2^2 \geqslant 0, \\ I_{2m-l}\sin(2m-l)\varphi, & k_2^2 < 0, \end{cases}$$

$$\widetilde{E}_2^{2i-1} \Leftrightarrow \begin{cases} N_{2m-l}\sin(2m-l)\varphi, & k_2^2 \geqslant 0, \\ K_{2m-l}\sin(2m-l)\varphi, & k_2^2 < 0, \end{cases}$$

$$H_2^{2i-1} \Leftrightarrow \begin{cases} J_{2m+l}\cos(2m+l)\varphi, & k_2^2 \geqslant 0, \\ I_{2m+l}\cos(2m+l)\varphi, & k_2^2 < 0, \end{cases} \tag{36}$$

$$\widetilde{H}_2^{2i-1} \Leftrightarrow \begin{cases} N_{2m+l}\cos(2m+l)\varphi, & k_2^2 \geqslant 0, \\ K_{2m+l}\cos(2m+l)\varphi, & k_2^2 < 0, \end{cases}$$

where the argument of the cylindrical functions is $\rho_i x_2$, $i = 1, 2$;

$$E_3 \Leftrightarrow K_{2m-l}\sin(2m-l)\varphi, \quad H_3 \Leftrightarrow K_{2m+l}\cos(2m+l)\varphi, \tag{37}$$

where the argument of the Macdonald functions is $\rho_2 x_3$ (in formulae (35)-(37) $m = 1, 2, \ldots$ with a $2m - l$ order of the functions and when $m = 0, 1$, with a $2m + 1$ order of the functions);

$$E_{\xi 1} \Leftrightarrow \frac{\epsilon_1 x_1}{k_1^2}\begin{cases} J_{2m-l-1}S_{2m-l} + J_{2m-l+1}\widetilde{S}_{2m-l}, & k_1^2 \geqslant 0, \\ I_{2m-l-1}S_{2m-l} - I_{2m-l+1}\widetilde{S}_{2m-l}, & k_1^2 < 0, \end{cases}$$
$$H_{\xi 1} \Leftrightarrow \frac{x_1}{k_1^2}\begin{cases} J_{2m+l-1}C_{2m+l} + J_{2m+l+1}\widetilde{C}_{2m+l}, & k_1^2 \geqslant 0, \\ I_{2m+l-1}C_{2m+l} - I_{2m+l+1}\widetilde{C}_{2m+l}, & k_1^2 < 0, \end{cases} \tag{38}$$

where the argument of the cylindrical functions is $\rho_1 x_1$, and the argument of the functions S, C is e_1, φ;

$$E_{\xi 2}^{2i-1} \Leftrightarrow \frac{x_2}{k_2^2}\begin{cases} J_{2m-l-1}S_{2m-l} + J_{2m-l+1}\widetilde{S}_{2m-l}, & k_2^2 \geqslant 0, \\ I_{2m-l-1}S_{2m-l} - I_{2m-l+1}\widetilde{S}_{2m-l}, & k_2^2 < 0, \end{cases}$$

$$\tilde{E}_{\xi 2}^{2i-1} \Leftrightarrow \frac{x_2}{k_2^2}\begin{cases} N_{2m-l-1} S_{2m-l} + N_{2m-l+1} \tilde{S}_{2m-l}, & k_2^2 \geqslant 0, \\ -K_{2m-l-1} S_{2m-l} + K_{2m-l+1} \tilde{S}_{2m-l}, & k_2^2 < 0, \end{cases}$$

$$H_{\xi 2}^{2i-1} \Leftrightarrow \frac{x_2}{k_2^2}\begin{cases} J_{2m+l-1} C_{2m+l} + J_{2m+l+1} \tilde{C}_{2m+l}, & k_2^2 \geqslant 0, \\ I_{2m+l-1} C_{2m+l} - I_{2m+l+1} \tilde{C}_{2m+l}, & k_2^2 < 0, \end{cases}$$

$$\tilde{H}_{\xi 2}^{2i-1} \Leftrightarrow \frac{x_2}{k_2^2}\begin{cases} N_{2m+l-1} C_{2m+l} + N_{2m+l+1} \tilde{C}_{2m+l}, & k_2^2 \geqslant 0, \\ -K_{2m+l-1} C_{2m+l} + K_{2m+l+1} \tilde{C}_{2m+l}, & k_2^2 < 0, \end{cases} \tag{39}$$

$$E_{\eta 2}^{2i-1} \Leftrightarrow U\left(\frac{x_2}{k_2^2} - \frac{x_2}{k_{2i-1}^2}\right)\begin{cases} J_{2m-l-1} C_{2m-l} - J_{2m-l+1} \tilde{C}_{2m-l}, & k_2^2 \geqslant 0, \\ I_{2m-l-1} C_{2m-l} + I_{2m-l+1} \tilde{C}_{2m-l}, & k_2^2 < 0, \end{cases}$$

$$\tilde{E}_{\eta 2}^{2i-1} \Leftrightarrow U\left(\frac{x_2}{k_2^2} - \frac{x_2}{k_{2i-1}^2}\right)\begin{cases} N_{2m-l-1} C_{2m-l} - N_{2m-l+1} \tilde{C}_{2m-l}, & k_2^2 \geqslant 0, \\ -K_{2m-l-1} C_{2m-l} - K_{2m-l+1} \tilde{C}_{2m-l}, & k_2^2 < 0, \end{cases}$$

$$H_{\eta 2}^{2i-1} \Leftrightarrow -U\left(\frac{x_2}{k_2^2} - \frac{x_2}{k_{2i-1}^2}\right)\begin{cases} J_{2m+l-1} S_{2m+l} - I_{2m+l+1} \tilde{S}_{2m+l}, & k_2^2 \geqslant 0, \\ I_{2m+l-1} S_{2m+l} + I_{2m+l+1} \tilde{S}_{2m+l}, & k_2^2 < 0, \end{cases}$$

$$\tilde{H}_{\eta 2}^{2i-1} \Leftrightarrow -U\left(\frac{x_2}{k_2^2} - \frac{x_2}{k_{2i-1}^2}\right)\begin{cases} N_{2m+l-1} S_{2m+l} - N_{2m+l+1} \tilde{S}_{2m+l}, & k_2^2 \geqslant 0, \\ -K_{2m+l-1} S_{2m+l} - K_{2m+l+1} \tilde{S}_{2m+l}, & k^2 < 0, \end{cases}$$

where the argument of the cylindrical functions is $\rho_i x_2$ and the argument of the S, C functions is e_i, φ, $i = 1, 2$;

$$E_{\xi 3} \Leftrightarrow -\frac{\epsilon_3 x_3}{k_3^2} \quad (K_{2m-l-1} S_{2m-l} - K_{2m-l+1} \tilde{S}_{2m-l}),$$

$$H_{\xi 3} \Leftrightarrow -\frac{x_3}{k_3^2} (K_{2m+l-1} C_{2m+l} - K_{2m+l+1} \tilde{C}_{2m+l}), \tag{40}$$

where the argument of the Macdonald functions is $\rho_2 x_3$, while the argument of the S and C functions are e_2, φ (in formulae (38)-(40) $m = 1, 2, \ldots$ with an order of the cylindrical functions equal to $2m - l \pm 1$ and $m = 0, 1, \ldots$ with an order of the cylindrical functions equal to $2m + l \pm 1$).

In formulae (35)-(40) l may take a value of either 0 or 1 depending on which even modes are considered.

In the case of odd modes it is necessary to carry out the following substitution of functions in (35)-(40): sin by cos and cos by sin, C by S, S by C as well as $m = 0, 1, 2, \ldots$ by $m = 1, 2, \ldots$, $m = 1, 2, \ldots$, by $m = 0, 1, 2, \ldots$ It is also necessary to change the sign of l and the sequences of functions determining the matrices

The following method was used to obtain the Fourier coefficients of the functions represented in (35)-(40).

Let there be a continuous periodic function $F(\cos^2\varphi)$ with period π. The following Fourier series expansion is valid for such a function

$$F(\cos^2\varphi) = \frac{1}{2}B_0 + \sum_{j=1}^{\infty} B_j \cos 2j\varphi .$$

We will determine the m values of the function $F(\cos^2\varphi)$ for $\varphi_i = 0.5\, i\pi/(m-1)$ for

$$f_i = F(\cos^2\varphi_i), \quad i = 0, 1, \ldots, m-1 .$$

We will draft an m-order square matrix A whose components are

$$a_{i1} = \frac{1}{m-1}, \quad a_{im} = \frac{(-1)^{i-1}}{m-1}, \quad i = 1, 2, \ldots, m-1,$$

$$a_{m1} = \frac{1}{2(m-1)}, \quad a_{mm} = \frac{(-1)^{m-1}}{2(m-1)}, \quad a_{mj} = \frac{(-1)^{j-1}}{m-1},$$

$$a_{ij} = \frac{2}{m-1} \cos\left[(i-1)(j-1)\frac{\pi}{m-1}\right], \quad a_{1j} = \frac{2}{m-1}, \quad \left\{\begin{matrix} i \\ j \end{matrix}\right\} = 2, 3, \ldots, m-1 .$$

We will determine the m-dimensional vector $\tilde{B} = (\tilde{B}_0, \tilde{B}_1, \ldots, \tilde{B}_{m-1})$ by the formula $\tilde{B} = Af$, where $f = (f_2, f_1, \ldots, f_{m-1})$, then for $m \to \infty$ $\tilde{B}_i \to B_i$, $i = 0, 1, \ldots, m-1$.

The following Fourier series expansion is valid for the functions $F_{sm}(\varphi) = F(\cos^2\varphi)\sin(2m - l)$, $m = 1, 2, \ldots$, $l = 0, 1$:

$$F_{sm}(\varphi) = \sum_{i=1}^{\infty} S_i \sin(2i - l),$$

where $S_i = 0{,}5\,(B_{|m-i|} - B_{|m+i-l|})$.

The following Fourier series is valid for the function $F_{cm}(\varphi) = F(\cos^2\varphi)\cos(2m + l)$, $m = 0, 1, \ldots$, $l = 0, 1$

$$F_{cm}(\varphi) = \sum_{i=0}^{\infty} C_i \cos(2i + l),$$

where $C_i = 0{,}5\,(B_{|m-i|} + B_{|m+i+l|})$.

Eliminating A_1, B_1, A_3, B_3 from equation system (33) we obtain a lower order equation system:

$$
\begin{aligned}
&[H^1_{\xi 2} - H_{\xi 1}(H_1)^{-1}H^1_2]\,\mathbf{B}_2 + [\tilde{H}^1_{\xi 2} - H_{\xi 1}(H_1)^{-1}\tilde{H}^1_2]\,\tilde{\mathbf{B}}_2 - E^1_{\eta 2}\,\mathbf{A}_2 - E^1_{\eta 2}\,\mathbf{A}_2 = 0, \\
&[E^1_{\xi 2} - E_{\xi 1}(E_1)^{-1}E^1_2]\,\mathbf{A}_2 + [\tilde{E}^1_{\xi 2} - E_{\xi 1}(E_1)^{-1}\tilde{E}^1_2]\,\tilde{\mathbf{A}}_2 + H^1_{\eta 2}\,\mathbf{B}_2 + \tilde{H}^1_{\eta 2}\,\tilde{\mathbf{B}}_2 = 0, \\
&[H^3_{\xi 2} - H_{\xi 3}(H_3)^{-1}H^3_2]\,\mathbf{B}_2 + [\tilde{H}^3_{\xi 2} - H_{\xi 3}(H_3)^{-1}\tilde{H}^3_2]\,\tilde{\mathbf{B}}_2 - E^3_{\eta 2}\,\mathbf{A}_2 - \tilde{E}^3_{\eta 2}\,\tilde{\mathbf{A}}_2 = 0, \\
&[E^3_{\xi 2} - E_{\xi 3}(E_3)^{-1}E^3_2]\,\mathbf{A}_2 + [\tilde{E}^3_{\xi 2} - E_{\xi 3}(E_3)^{-1}\tilde{E}^3_2]\,\tilde{\mathbf{A}}_2 + H^3_{\eta 2}\,\mathbf{B}_2 + \tilde{H}^3_{\eta 2}\,\tilde{\mathbf{B}}_2 = 0.
\end{aligned}
\tag{41}
$$

This is a homogeneous system of equations with constant coefficients and unknown $\mathbf{A}_2$, $\mathbf{B}_2$, $\tilde{\mathbf{A}}_2$, $\tilde{\mathbf{B}}_2$. In order for this system of equations to have a nontrivial solution, its determinant d must be equal to zero. Substituting these expressions (12), (15) through U and V in place of $\varkappa_{1,2,3}$, $k^2_{1,2,3}$, we obtain $D(U, V) = 0$. This is the characteristic equation which in implicit form determines the dependence of U on V.

4. Calculation Results for the Dispersion Characteristics of Elliptical Optical Fibers

Using the method proposed above for calculating three-layer elliptical optical fibers a computer calculation was performed for the following characteristics of optical fibers for the HE^x_{11} odd (working) mode in which the vector **E** lies on the semi-major elliptical axis (see Fig. 1), and for the HE^y_{11} orthogonally-polarized mode:

- the dimensionless transverse wave numbers $\varkappa_{1,2,3}$ in the core, the intermediate and outer claddings of the optical fiber (see (12));

- the phase velocity retardation coefficient $u = \upsilon_2/\upsilon_{\text{п}}$, where $\upsilon_2 = c/n_2$;

- The group velocity retardation coefficient $U_{\text{гр}} = \upsilon_2/\upsilon_{\text{гр}}$ and its derivative with respect to V, where V is the characteristic parameter of the optical fiber;

- The difference in the retardations of the phase and group velocities of the odd and even modes.

All these characteristics were calculated for step refractive index optical fibers, specifically:

- for optical fibers with an optically denser core in which $n_1 > n_2 = n_3$ or $\varepsilon_1 > \varepsilon_3 = 1$ (dual-layer optical fibers), $n_1 > n_3 > n_2$ or $\varepsilon_1 > \varepsilon_3 > 1$ (W-optical fibers), $n_1 > n_2 > n_3$ or $\varepsilon_1 > 1 > \varepsilon_3$;

- for circular waveguides with an optically denser intermediate cladding in which $n_2 > n_1 = n_3$ for $1 > \varepsilon_1 = \varepsilon_3$, $n_2 > n_1 > n_3$ or $1 > \varepsilon_1 > \varepsilon_3$, $n_2 > n_3 > n_1$ for $1 > \varepsilon_3 > \varepsilon_1$.

We will employ normalized coefficients of the phase and group mode retardations for ease of representation of the derived results and to make it possible to use these results to evaluate the characteristics of analogous optical fibers having other parameters ($n_{1,2,3}$, a, b) and different working wavelengths:

$$B = \frac{n_3^2 - n_2^2}{|n_1^2 - n_2^2|} = \frac{U^2 - 1}{|\epsilon_1 - 1|} \approx \frac{U - 1}{|\sqrt{\epsilon_1} - 1|}, \tag{42}$$

$$B_{\text{гр}} = \frac{n_{\text{гр}}^2 - n_2^2}{|n_1^2 - n_2^2|} = \frac{U_{\text{гр}}^2 - 1}{|\epsilon_1 - 1|} \approx \frac{U_{\text{гр}} - 1}{|\sqrt{\epsilon_1} - 1|}, \tag{43}$$

where $n_\varepsilon = c/v_\Phi$; $n_{\text{гр}} = c/v_{\text{гр}}$ are the retardation factors of the phase and group mode velocities; $U = n_\varepsilon/n_2$, $U_{\text{гр}} = n_{\text{гр}}/n_2$ are the retardation coefficients of the phase and group mode velocities; $\varepsilon_1 = n_1^2/n_2^2$, $\varepsilon_3 = n_3^2/n_2^2$. When $|n_1 - n_2| \ll 1$, $|n_2 - n_3| \ll 1$ the relative differentials of the refractive indices of the core and the intermediate cladding are

$$\Delta = |n_1 - n_2|/n_2 \approx |\epsilon_1 - 1|/2. \tag{44}$$

and the same values for the intermediate and outer claddings are

$$\Delta' = |n_2 - n_3|/n_2 \approx |1 - \epsilon_3|/2. \tag{45}$$

All optical fiber characteristics were determined as a function of the optical fiber parameter

$$V = 2\pi b_1 \lambda^{-1} |n_1^2 - n_2^2|^{1/2} = 2\pi b_1 \lambda_2^{-1} |\epsilon_1 - 1|^{1/2}, \tag{46}$$

where $\lambda_2 = \lambda/n_2$ with fixed eccentricities

$$e_1 = (1 - b_1^2/a_1^2)^{1/2}, e_2 = (1 - b_2^2/a_2^2)^{1/2} \tag{47}$$

and a fixed ratio of the semi-minor elliptical axes of the layer interfaces

$$c = b_1/b_2. \tag{48}$$

It is easily demonstrated that the normalized retardation coefficients of the phase and group mode velocities are related by the relation

$$B_{\text{гр}} = dVB/dV. \tag{49}$$

In the first approximation for optical fibers with uncompensated dispersion, the broadening of a pulse having a spectral linewidth $\Delta\lambda$ and transiting a optical fiber of length L is

$$\Delta\tau = S_1 L \Delta\lambda, \tag{50}$$

where S_1 is the first order dispersion coefficient [5], which when $\Delta n = n_1 - n_2 \ll 1$ is equal to

$$S_1 = -\frac{1}{\lambda c}\left(\lambda^2 \frac{d^2 n_2}{d\lambda^2} + \Delta n V \frac{d^2 VB}{dV^2}\right),$$

i.e., is determined by the sum of the material and waveguide dispersions.

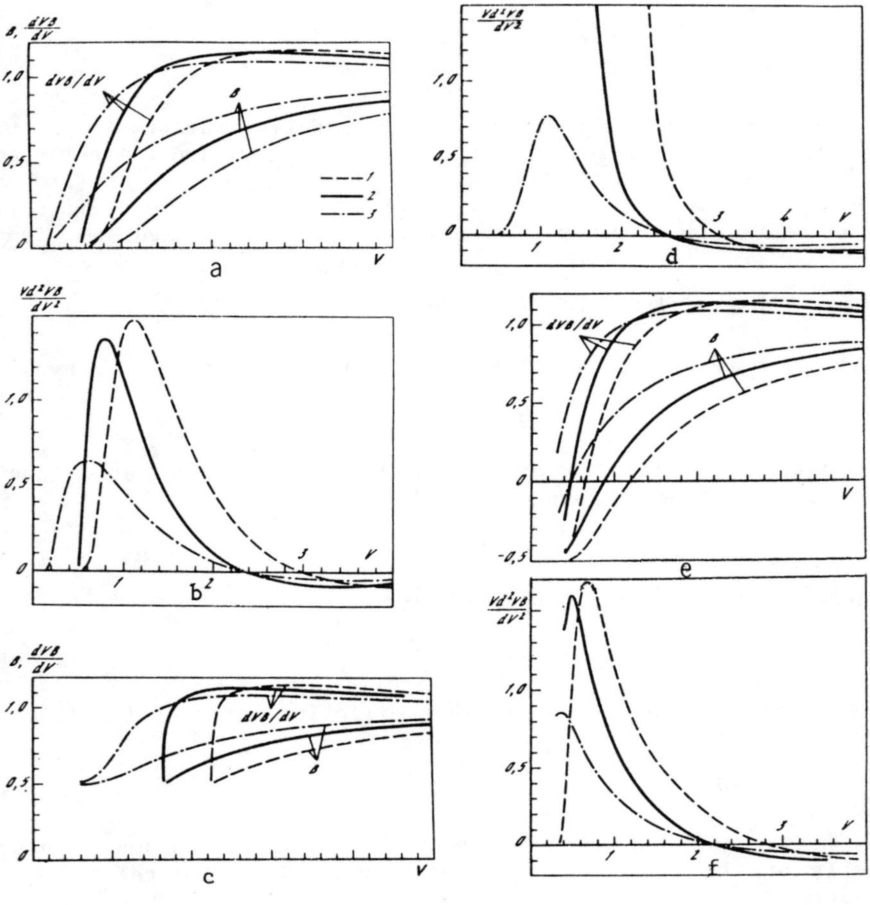

Fig. 3. The normalized phase and group retardation coefficients (a, c, d) and the waveguide dispersion (b, d, f) plotted as a function of V for the HE^{x}_{11} mode of a optical fiber with $n_1 > n_2 = n_3$ (a, b), $n_1 > n_3 > n_2$ (c, d), and $n_1 > n_2 > n_3$ (e, f)
1 - e = 0; 2) $e = \sqrt{3/2}$; 3) 3 = 1

The quantity $\lambda^2 d^2 n_2/d\lambda^2$ characterizing the material dispersion is sufficiently fully determined for optical fiber materials based on silica and boron-, germanium-, and phosphor-doped silica [5].

The waveguide dispersion $\Delta n V d^2 VB/dV^2$ has not yet been calculated for elliptical optical fibers. Therefore the quantity $V d^2 VB/dV^2$ was calculated making it possible to find the waveguide dispersion for optical fibers with a different Δn.

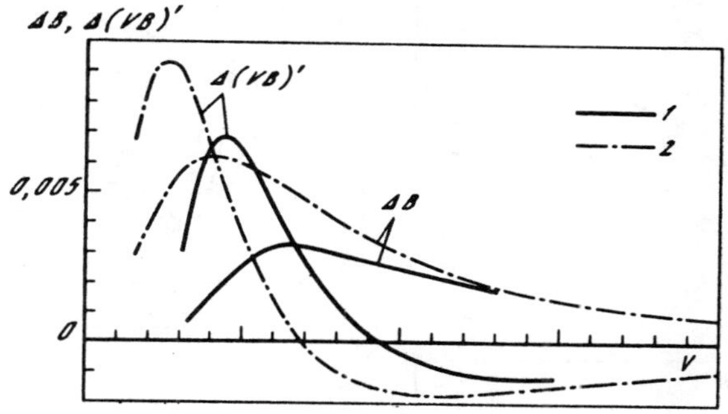

a

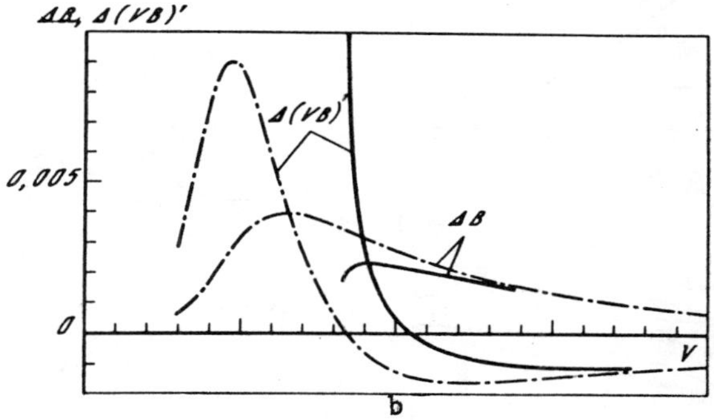

b

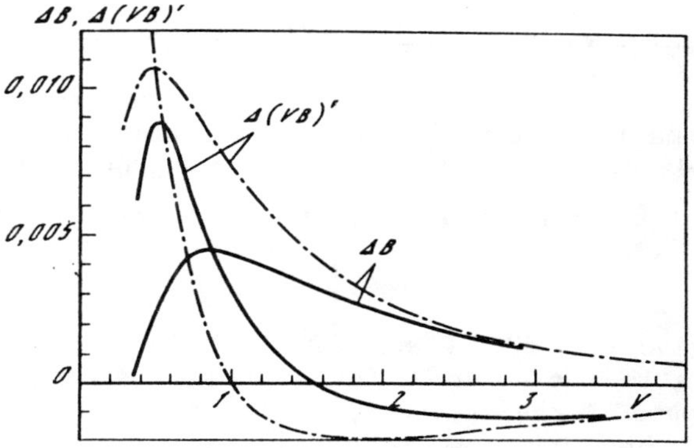

c

Fig. 4. The differences of the normalized phase and group retardation coefficients of the HE_{11}^{x} and HE_{11}^{y} modes plotted against V for optical fibers with $n_1 > n_2 = n_3$ (a), $n_1 > n_3 > n_2$ (b) and $n_1 > n_2 > n_3$ (c) 1) $e = \sqrt{3}/2$; 2) $e = 1$

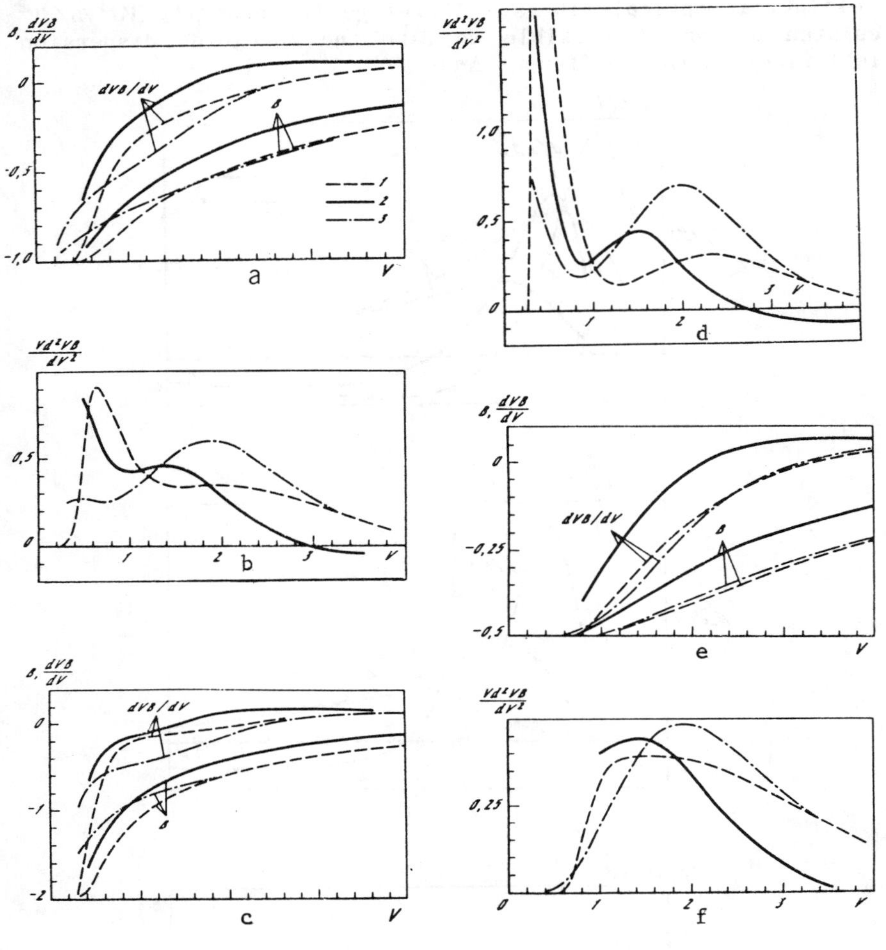

Fig. 5. The normalized phase and group retardation coefficients (a, c, d) and waveguide dispersion (b, d, f) plotted as a function of V for the HE^{x}_{11} mode of a circular optical fiber with $n_2 > n_1 = n_3$ (a, b), $n_3 > n_1 > n_3$ (c, d), and $n_2 > n_3 > n_1$ (e, f)
1) $e = 0$; 2) $e = \sqrt{3}/2$; 3) $e = 1$

Fig. 3 gives B, dVB/dV and Vd^2VB/dV^2 plotted as a function of the characteristic parameter V for identical eccentricities of the interface ellipses of the layers $e_1 = e_2 = e = \sqrt{3}/2$ (i.e., for $a_1/b_2 = a_2/b_2 = 2$) and identical ratios $c = b_1/b_2 = 0.5$ for optical fibers with a denser core, for which $n_1 > n_2 = n_3$ or $\varepsilon_1 > \varepsilon_3 = 1$ (Fig. 3a, b), $n_1 > n_3 > n_2$ or $\varepsilon_1 > \varepsilon_3 > 1$ (Fig. 3c, d), $n_1 > n_2 > n_3$ or $\varepsilon_1 > 1 > \varepsilon_3$ (Fig. 3e, f). In the same figure the dashed and dot-dash curves represent the same relations for analogous optical fibers of circular cross-section ($e = 0$) and form plane five-layer optical fibers ($e = 1$), i.e., relations corresponding to the minimum and maximum cross-sectional ellipticities of the optical fibers. Fig. 3 gives the relations for the HE_{11}^{x} odd modes which virtually coincide with analogous relations for the HE_{11}^{y} even modes since they were calculated for optical fibers with minor differences in the refractive indices of the layers.

Fig. 4 demonstrates how the phase and group velocities of the odd and even HE modes are differentiated, where $\Delta B = B_{HE_{11}^{x}} - B_{HE_{11}^{y}}$, $\Delta(VB)' = (dVB/dV)_{HE_{11}^{x}} - (dVB/dV)_{HE_{11}^{y}}$. It is clear from an analysis of these relations that there exist optimum values of V at which the difference in the phase velocities is maximized and, consequently, in this case the transformation of these modes by microbends will be minimized. We should also note that there exist values of B where $\Delta(VB)' = 0$, i.e., the group velocities of the orthogonally-polarized modes are identical.

Fig. 5 provides analogous relations for circular waveguides in which $n_2 > n_1 = n_3$ or $1 > \varepsilon_1 = \varepsilon_3$ (Fig. 5a, b), $n_2 > n_1 > n_3$ or $1 > \varepsilon_1 > \varepsilon_3$ (Fig. 5c, d), $n_2 > n_3 > n_1$ or $1 > \varepsilon_3 > \varepsilon_1$ (Fig. 5e, f).

Conclusion

The proposed method has made it possible to calculate the dispersion characteristics of three-layer elliptical optical fibers which makes it possible to more exactly account for the influences of various optical fiber parameters on signal transmission in single-polarization, single-mode as well as dual-mode operation and on the conversion of orthogonally-polarized modes.

The method employed here also makes it possible to calculate the dispersion characteristics of optical fibers with various eccentricities of the layer interface ellipses and with different relative orientations of the semiaxes of these ellipses.

Bibliography

1. Belanov, A.S., Vzyatyshev, V.F. "The distribution of a longitudinal energy flux in a circular dielectric waveguide" Moscow: VZMI, 1969, p. 12-22.

2. Belanov, A.S., Dianov, Ye.M., Ezhov, G.I., Prokhorov, A.M. "The propagation of natural waves in multilayer optical waveguides. Volumes 1 and II" KVANTOVAYA ELEKTRON, 1976, V. 3, No. 1, p. 81-93.

3. Belanov, A.S., Dianov, Ye.M., Prokhorov, A.M. "Three-layer wave-

guides for broadband optical communications links" IZV. AN SSSR, SER. FIZ, 1978, V. 42, No. 12, p. 2522-2533.

4. Belanov, A.S., Dianov, Ye.M., Prokhorov, A.M. "Data transmission on quasi single-mode three-layer optical waveguides" KVANTOVAYA ELEKTRON., 1979, V. 6, No. 12, p. 2646-2649.

5. Belanov, A.S., Dianov, Ye.M. "Maximum data transmission speeds on optical fibers" RADIOTEKHNIKA, 1982, V. 37, No. 2, p. 35-43.

6. Belanov, A.S., Dianov, Ye.M., Krivenkov, V.I. "optical fibers with stable emission polarization" KVANTOVAYA ELEKTRON., 1984, V. 11, No. 6, p. 1273-1275.

7. Lyubimov, L.A., Beselov, G.I., Bey, N.A. "Dielectric waveguides of elliptical cross-section" RE, 1961, V. 6, No. 11, p. 1871-1880.

8. Yeh, C.W. "On the elliptical dielectric waveguides" J. APPL. PHYS., 1962, V. 33, p. 3235-3243.

9. Yeh, C.W. "Attenuation in a dielectric elliptical cylinder" IEEE TRANS. ANTENNAS and PROPAG. 1963, V. Ap-II, No. 2, p. 177-184.

10. Hondros, D., Debye, P. "Elektromagnetische Wellen an dielektrischen Drahten" ANN. PHYS., 1910, V. 32, No. 8, p. 465-476.

11. Yeh, C. "Modes in weakly guiding elliptical optical fibers" OPT. AND QUANT. ELECTRON., 1976, V. 8, No. 1, p. 43-47.

12. Rengarajan, S.R., Lewis, Z.E. "Single-mode propagation in multilayer elliptical fiber waveguides" RADIO SCI., 1981, V. 16, No. 4, p. 541-547.

The Coherence and Polarization Properties of Emission in Single-Mode Optical Fibers

A.B. Grudinin, V.B. Sulimov

Abstract: This chapter consists of a theoretical and experimental section. The first section considers the change in the coherence and polarization characteristics of quasi-monochromatic radiation propagating through a monomode optical fiber when accounting for the dispersion of optical fiber parameters. The spectral density of the Stokes vector is introduced and an equation is derived for the spectral density in an inhomogeneous monomode optical fiber. This equation is analyzed to establish conditions where the radiation at the optical fiber exit is completely depolarized independent of the polarization state of the injection radiation. The second half of this chapter presents an experimental technique and results from measurement of the radiation correlation function at the exit of low and high birefringence optical fibers. The spectral dependencies of first and second order polarization dispersion are derived for low and high birefringence optical fibers, respectively. It is demonstrated that theory and experiment are in agreement.

Introduction

From the viewpoint of data transmission systems single-mode optical fibers based on silica glass represent a near-ideal transmission medium, since their optical losses amount to fractions of a decibel per kilometer in the near IR, and the primary factor limiting the bandwidth of single-mode optical fibers is intramodal dispersion which may be reduced to zero by proper selection of the optical fiber parameters.

One of the fundamental properties of the HE_{11} fundamental mode of optical fibers is that it may exist in two orthogonal polarization states and in the general case the propagation constants are different for orthogonally polarized modes, i.e., in reality a single-mode optical fiber is a two-mode optical fiber. The degree of polarization and the polarization state of a wave at the exit of such a single-mode optical fiber will be determined primarily by the difference in the propagation constants of the naturally-polarized modes of the optical fiber and the coherence properties of the emission source. Since the difference of the propagation constants additionally depend on external influences on the optical fiber, the variation in the polarization state of the wave at the exit of the optical fiber may be used to determine their magnitude which in conjunction with the spatial field coherence in the optical fiber makes it possible to fabricate high-sensitivity transducers.

Investigations carried out both domestically and abroad have revealed that depending on the specific problem involving the polarization properties of emission, an optical fiber will either leave unchanged the polarization state and polarization degree of emission or will completely depolarize such emission. Since quasi-monochromatic emission may be depolarized, the problem is in fact reduced to considering the influence of the coherence properties of emission on the polarization properties of an optical fiber. The solution to this problem is important not only for the successful design of fiber-optic sensors but also from the viewpoint of the application of single-mode optical fibers in high-speed data transmission systems, since the difference in the propagation constants of the eigenmodes of the optical fiber produce polarization dispersion that may limit the bandwidth of the system.

In this regard it is advisable to investigate both theoretically and experimentally the coherence and polarization properties of quasi-monochromatic emission propagating in single-mode optical fibers.

1. Theory

As we know [1-3] the fundamental HE_{11} mode of an ideal circular weakly-guiding dielectric waveguide is linearly polarized and may exist in two orthogonal states, with equal propagation constants of these eigenmodes in such an isotropic optical fiber (which clearly results from the circular symmetry of the optical fiber cross-section). If the dielectric constant of the waveguiding layers in a single-mode optical fiber is isotropic, while the profile is constant on its axis (the oz axis) and has two orthogonal planes of symmetry passing through o_x and o_y axes (henceforth the field components on these axes will be labeled by 1 and 2, respectively), the eigenmodes of such an optical fiber are also linearly-polarized with their planes of polarization coinciding with the axes of symmetry, although the propagation constants of these modes differ due to the variable electromagnetic wave propagation through the optical fiber layers. We will call such an optical fiber an ideal anisotropic single-mode optical fiber.

In an actual optical fiber various inhomogeneities and strains will produce interaction of the linearly-polarized modes. In this case the Fourier components of the transverse field components may be expanded in terms of the transverse eigenfunctions of an ideal optical fiber; the coefficients of this expansion E_1 and E_2 (for an electric field) will satisfy the following system of coupled mode equations [3, 4]:

$$\frac{dE(\omega,z)}{dz} = TE(\omega,z), \tag{1}$$

where

$$E = \begin{Vmatrix} E_1 \\ E_2 \end{Vmatrix}; \quad T = \begin{Vmatrix} -i\beta_1 & h \\ -h^* & -i\beta_2 \end{Vmatrix}. \tag{2}$$

Here β_1, β_2 are the propagation constants of the eigenmodes of an ideal optical fiber (neglecting optical fiber losses and we will take them to be real); h is the intermodal coupling coefficient which is generally complex, $h = |h|\exp(i\sigma)$ (henceforth we shall designate $|h|$ simply as h where it will not cause confusion), while the * infers complex conjugation.

With internal strains produced by the elastooptic effect an anisotropic addition to the refractive index will arise which may not only result in a corresponding contribution to the intermodal coupling coefficient h but also to the propagation constants β_1 and β_2 [5]. Generally speaking the parameters β_1, β_2, h and σ may change along the optical fiber and clearly the most significant limitation on the characteristic length of such changes derives from the requirement for zero losses in the optical fiber ($\mathrm{Im}\beta_1 = \mathrm{Im}\beta_2 = 0$).

Uniform optical fiber. This case when the parameters β_1, β_2, h and σ are independent of z is the simplest case and therefore we will consider it first. Equation (1) is easily solved if we know the eigenvectors and the eigenvalues of matrix T. We may easily find the eigenvalues

$$\lambda_{1,2} = i(-\beta \pm \gamma) \tag{3}$$

and their orthonormalized eigenvectors

$$\mathcal{E}_{1,2} = \frac{1}{\sqrt{2}} \begin{Vmatrix} (1 \mp \kappa)^{1/2} \\ e^{-i(\sigma \mp \pi/2)}(1 \pm \kappa)^{1/2} \end{Vmatrix}. \tag{4}$$

Here the superscript and subscript are taken for the first and second eigenvalues (vectors) respectively, and we introduce the following conventions:

$$\gamma = [\,|h|^2 + (\Delta\beta/2)^2\,]^{1/2},$$
$$\beta = (\beta_1 + \beta_2)/2, \Delta\beta = \beta_1 - \beta_2, \kappa = \Delta\beta/2\gamma.$$

Obviously vectors (4) correspond to the monochromatic waves

$$\exp\{\, i[\omega t - (\beta \mp \gamma)z]\}\ \mathcal{E}_{1,2},$$

whose polarization ellipses are identical and are configured so that the directions of their semi-major axes are orthogonal, while the directions of revolution of the field vectors in both modes are opposed. These represent the eigenmodes of a real anisotropic single-mode optical fiber when coupling arises between the eigenmodes of the ideal anisotropic single-mode optical fiber. The propagation constants of the eigenmodes of an actual anisotropic single-mode optical fiber are equal ($\beta \mp \gamma$) and the semiaxes of the polarization ellipses are given by the expression

$$|(1+\chi\cos\sigma)^{1/2} \pm (1-\chi\cos\sigma)^{1/2}|,$$

with the angle θ_1 between the direction of the semi-major axis of the ellipse of the first mode (whose field vector rotation is positive for $|\sigma| < \pi/2$) and the ox axis is equal to

$$\theta_1 = -\frac{1}{2}\,\mathrm{arctg}\left(\frac{\chi}{\kappa}\sin\sigma\right),$$

where $\chi = (1-\kappa^2)^{1/2} = |h|/\gamma$.

It is clear from the expressions in (4) that only when $\sigma = \pm\pi/2$ when the intermodal coupling coefficient is an imaginary quantity will both eigenmodes of the optical fiber be linearly-polarized (this corresponds to the case of linear birefringence). In the case $\mathrm{Re}h \neq 0$ the eigenmodes are elliptically polarized, which indicates optical activity in the optical fiber. We note that by rotating the coordinate system about the optical fiber axis by an angle

$$\mu = \frac{1}{2}\,\mathrm{arctg}\left(\frac{\Delta\beta}{2|h|\sin\sigma}\right)$$

matrix T in equation (1) may be reduced to a form with identical diagonal elements $-i\beta$, although with a different intermodal coupling coefficient

$$h' = |h|\cos\sigma + i\,[(\Delta\beta/2)^2 + (|h|\sin\sigma)^2]^{1/2}.$$

However we should emphasize that the physical mechanisms resulting in $\beta_1 \neq \beta_2$ and mode coupling may be different and this produces different spectral relations $\Delta\beta(\omega)$ and $h(\omega)$ and, consequently, a different frequency dependence of the angle of rotation of the $\mu(\omega)$ coordinate axes. Thus, in examining the frequency set we must consider the general case where $\Delta\beta \neq 0$.

Then we find the solution of equation (1)

$$E(\omega,z) = F(\omega,z)\,E(\omega,0), \tag{5}$$

satisfying the boundary condition $E(\omega,z=0) = E(\omega,0) = \begin{Vmatrix} E_1(\omega,0) \\ E_2(\omega,0) \end{Vmatrix}$.

Here the unitary propagation matrix F takes the form ($\eta = \gamma z$)

$$F(\omega, z) = e^{-i\beta z} \begin{Vmatrix} \cos\eta - i\kappa \sin\eta & \chi e^{i\sigma} \sin\eta \\ -\chi e^{-i\sigma} \sin\eta & \cos\eta + i\kappa \sin\eta \end{Vmatrix}. \quad (6)$$

Making the transformation to functions of time we obtain from relation (5)

$$E(t, z) = \frac{1}{2\pi} \int dt' F(t', z) E(t - t', 0), \quad (7)$$

where $F(t, z)$ and $E(t, z)$ are the transforms of the Fourier functions of the frequencies $F(\omega, z)$ and $E(\omega, z)$.

We will consider the case of quasi-monochromatic emission where the field at the optical fiber exit takes the form

$$E(t, 0) = \bar{E}(t, 0) e^{i\omega_0 t}, \quad (8)$$

where $\bar{E}(t, 0)$ is a slowly varying (compared to the rapidly oscillating factor $\exp(i\omega_0 t)$) function of time; ω_0 is the optical frequency. Substituting expression (8) into relation (7), we obtain for the emission amplitude at the optical fiber exit:

$$\bar{E}(t, z) = \frac{1}{2\pi} \int dt' e^{-i\omega_0 t'} F(t', z) \bar{E}(t - t', 0). \quad (9)$$

In order to analyze the polarization properties of emission traveling along the optical fiber we will introduce the correlation tensor whose components are equal to

$$J_{ij}(t - t', z) = \langle \bar{E}_i(t, z) \bar{E}_j^*(t', z) \rangle, \quad (10)$$

where <...> infers time averaging (we assume that the complex amplitudes of the emission components $\bar{E}(t)$ are steady-state random processes). For $t = t'$ $J(0, z)$ is the familiar coherency matrix (6) whose components determine the polarization state and the emission intensity. Making the transformation to the spectral density of the correlation tensor $\tilde{J}(\omega, z)$, determined by the relation

$$J(t - t', z) = \int d\omega \exp[i\omega(t - t')] \tilde{J}(\omega, z), \quad (11)$$

and substituting (9) into (10) subject to the stationarity of the emission we obtain

$$\tilde{J}(\omega, z) = F(\omega_0 + \omega, z) \tilde{J}(\omega, 0) F^+(\omega_0 + \omega, z), \quad (12)$$

where F^+ is the Hermitian conjugate matrix to matrix F.

It is easily determined that in accordance with formulae (11) and (12) due to the unitarity of matrix F the spur of the coherency matrix determining the total emission intensity does not change (which is not surprising since we are neglecting optical fiber losses) and therefore the correlation tensor may easily be normalized to the spur

$$\Gamma(\tau, z) = J(\tau, z)/\mathrm{Sp}J(0, z) \tag{13}$$

and the coherency matrix

$$\rho(z) = \Gamma(0, z) = J(0, z)/\mathrm{Sp}J(0, z), \tag{14}$$

which in this case is normally called the polarization tensor [7]. It is clear from formula (12) that the spectral density spur of the correlation tensor is conserved along the optical fiber for each frequency component (as would be expected in the absence of nonlinear processes) and, consequently, the spur of the correlation tensor itself is conserved for each τ.

It is convenient to employ the Stokes parameters ξ_1, ξ_2, ξ_3 to characterize the polarization properties of emission; these parameters may easily be related to the polarization tensor components [7]:

$$\xi_1 = 2\mathrm{Re}\rho_{12}, \quad \xi_2 = -2\mathrm{Im}\rho_{12}, \quad \xi_3 = \rho_{11} - \rho_{22}. \tag{15}$$

The primary convenience of these parameters derives from the fact that they are real and are physically measurable quantities. The quantities ξ_1, ξ_2, ξ_3 may easily be represented as the components of a certain vector in special three-dimensional space called the Stokes space $\boldsymbol{\xi} = (\xi_1, \xi_2, \xi_3)$ [6].

We will discuss certain properties of the correlation tensor introduced above. Using definitions (10) and (13) we may easily show that

$$\Gamma^*_{ij}(\tau) = \Gamma_{ji}(-\tau). \tag{16}$$

From here we obtain the following relation for the spectral density components of the correlation tensor:

$$\Gamma^*_{ij}(\omega) = \Gamma_{ji}(\omega). \tag{17}$$

It is clear that the components of polarization tensor (14) also satisfy relations (17) and this indicates the possibility of introducing three real quantities similar to Stokes parameters for each frequency component of the correlation tensor. By definition we incorporate the quantities [8]

$$\zeta_1(\omega) = 2\mathrm{Re}\Gamma_{12}(\omega), \quad \zeta_2(\omega) = -2\mathrm{Im}\Gamma_{12}(\omega),$$
$$\zeta_3(\omega) = \Gamma_{11}(\omega) - \Gamma_{22}(\omega), \tag{18}$$

which, like Stokes parameters, form the real vector $\zeta(\omega)$ which represents the spectral density of the Stokes vector. Indeed we may easily obtain by means of relations (11), (13)-(15), and (17):

$$\boldsymbol{\xi} = \int d\omega \boldsymbol{\zeta}(\omega). \tag{19}$$

Therefore the quantities

$$\boldsymbol{\zeta}(\tau) = \int d\omega e^{i\omega\tau} \boldsymbol{\zeta}(\omega), \tag{20}$$

which may be called the correlation Stokes parameters represent a generalization of the regular Stokes parameters and are related to the components of the correlation tensor by the relations

$$\begin{aligned} \zeta_1(\tau) &= \Gamma_{12}(\tau) + \Gamma_{21}(\tau), \\ \zeta_2(\tau) &= i[\Gamma_{12}(\tau) - \Gamma_{21}(\tau)], \\ \zeta_3(\tau) &= \Gamma_{11}(\tau) - \Gamma_{22}(\tau). \end{aligned} \tag{21}$$

We easily obtain from relation (16)

$$\boldsymbol{\zeta}^*(\tau) = \boldsymbol{\zeta}(-\tau). \tag{22}$$

We will now find the expression for the Stokes parameters of the emission at the optical fiber exit accounting for the frequency dependence of the propagation constants β_1, β_2 and the intermodal coupling parameter h. We may easily find expressions for the Fourier components of the correlation tensor at the optical fiber exit by means of formulae (6), (12), and (13) and for Γ_{11} and Γ_{12} they take the form

$$\begin{aligned} \Gamma_{11}(\omega, z) &= \Gamma_{11}(\omega, 0)(\cos^2\eta + \kappa^2\sin^2\eta) + \Gamma_{22}(\omega, 0)\chi^2\sin^2\eta + \\ &+ 2\mathrm{Re}[\Gamma_{12}(\omega, 0)(\cos\eta - i\kappa\sin\eta)e^{-i\sigma}]\chi\sin\eta, \end{aligned} \tag{23}$$

$$\begin{aligned} \Gamma_{12}(\omega, z) &= [-\Gamma_{11}(\omega, 0) + \Gamma_{22}(\omega, 0)]e^{i\sigma}\chi\sin\eta(\cos\eta - i\kappa\sin\eta) + \\ &+ \Gamma_{12}(\omega, 0)(\cos\eta - i\kappa\sin\eta)^2 - \Gamma_{21}(\omega, 0)e^{2i\sigma}\chi^2\sin^2\eta. \end{aligned} \tag{24}$$

In these expressions η, κ, χ, and σ are functions of $(\omega_0 + \omega)$. The expressions for the other components of the correlation tensor may easily be found from these formulae by means of relations (17) and equality

$$\mathrm{Sp}\,\Gamma(\omega, z) = \mathrm{Sp}\,\Gamma(\omega, 0). \tag{25}$$

We may then take the following inequalities as satisfied:

$$\left|\frac{d\sigma}{d\omega}\right|\Delta\omega \ll 1, \quad \left|\frac{d\ln\chi}{d\omega}\right|\Delta\omega \ll 1, \quad \left|\frac{d\ln\kappa}{d\omega}\right|\Delta\omega \ll 1, \tag{26}$$

$$\left|\frac{d^2\eta}{d\omega^2}\right|^{1/2}\Delta\omega \ll 1, \tag{27}$$

where $\Delta\omega$ is the spectral width of the emission source and all quantities are taken at frequency ω_0. The first condition in (26) allows us to assume $\Gamma(2\sigma', 0) \simeq \Gamma(0, 0)$, while the remaining two conditions in (26) make it possible to neglect the frequency dependence of all type χ, κ, etc. multipliers. Physically this infers an assumption that the polarization state of the eigenmode does not change in the interval $\Delta\omega$. When condition (27) is satisfied we may be limited to linear terms in the expansion of $\eta(\omega)$ in powers of ω, i.e., we may neglect second order polarization dispersion. For example, condition (27) is satisfied for $\Delta\omega/\omega \sim 10^{-3}$ and for an optical fiber with $\Delta\beta \sim 10\ \mathrm{cm}^{-3}$ at a length of $z < 10$ km. Condition (27) may be violated for either a sufficiently long optical fiber or one with a high birefringence. This case will be examined below. Assuming $\eta(\omega_0 + \omega) = \eta(\omega_0) + \eta'(\omega_0)\omega$, $\kappa(\omega_0 + \omega) = \kappa(\omega_0)$, $\chi(\omega_0 + \omega) = \chi(\omega_0)$, $\sigma(\omega_0 + \omega) = \sigma(\omega_0)$, in expressions (23), (24) and after integrating them over frequency ω, we obtain, subject to relations (15), (18), and (21),

$$\xi_1(z) = \chi A \sin\sigma + (\kappa B \sin\sigma + C\cos\sigma)\cos 2\eta + (\kappa C \sin\sigma - B\cos\sigma)\sin 2\eta, \tag{28}$$

$$\xi_2(z) = \chi A \cos\sigma + (\kappa B \cos\sigma - C\sin\sigma)\cos 2\eta + (\kappa C\cos\sigma + B\sin\sigma)\sin 2\eta, \tag{29}$$

$$\xi_3(\omega) = -\kappa A + \chi B\cos 2\eta + \chi C\sin 2\eta. \tag{30}$$

Here we introduce the conventions:

$$A = -(\mathbf{w}, \xi(0)), \tag{31}$$

$$B = \mathrm{Re}\,M, \quad C = -\mathrm{Im}\,M, \tag{32}$$

$$M = (\mathbf{u} - i\mathbf{v}, \zeta(\tau_1, 0)), \tag{33}$$

where **w**, **u**, and **v** are perpendicular unit vectors in the Stokes space:

$$\mathbf{w} = (-\chi\sin\sigma,\ -\chi\cos\sigma,\ \kappa), \tag{34}$$

$$\mathbf{u} = (\kappa\sin\sigma,\ \kappa\cos\sigma,\ \chi), \tag{35}$$

$$\mathbf{v} = (\cos\sigma,\ -\sin\sigma,\ 0), \tag{36}$$

where $\tau_1 = 2d\eta(\omega_0)/d\omega_0 = 2zd\gamma(\omega_0)/d\omega_0$ is the first order polarization dispersion.

The meaning of the unit vectors given above is clear. The vector **w** clearly corresponds to the excitation of one of the optical fiber eigenmodes, which may easily be determined directly from expression (4) and definitions (10), (14), and (15). The direction of vector **v** coincides with the line of intersection of the plane $\xi_1 o \xi_2$ in the Stokes space with the plane perpendicular to the axis of the eigenmodes determined by vector **w**. Obviously the vector **u** also lies in this plane and is perpendicular to both vectors **w** and **v**.

The degree of polarization P of the emission is expressed through the Stokes parameters:

$$P = (\xi_1^2 + \xi_2^2 + \xi_3^2)^{1/2}. \tag{37}$$

From here by means of expressions (28)–(30) we obtain

$$P = (A^2 + |M|^2)^{1/2}. \tag{38}$$

As would be expected the degree of emission polarization depends on the initial polarization state (through the parameters A and M) as well as the coherence properties of the emission (through the parameter M). Since with sufficient optical fiber length the delay between its two eigenmodes will begin to exceed the coherence time of the emission, the correlation Stokes vector $\zeta(\tau_1)$ vanishes, while the degree of polarization is equal to $|A|$, i.e., the value of the projection of the Stokes vector of the input emission onto the axis of the eigenmodes. Thus, the parameter M characterizes the mutual coherence of the optical fiber eigenmodes. In accordance with formulae (28)-(30) the Stokes vector diagram for the case of emission propagating along the optical fiber appears as a spiral lying in the plane perpendicular to the eigenmode axis and spiraling towards this axis in the limit $\tau_1 \gg \tau_c$, where τ_c is the coherence time of the emission source, where $\zeta(\tau_1) = 0$.

The following concepts clearly reveal the physical meaning of parameter A. We will expand the optical fiber field in terms of the eigenmode fields

$$E(z) = a_1 \mathcal{E}_1 \exp(-i\beta z + i\eta) + a_2 \mathcal{E}_2 \exp(-i\beta z - i\eta),$$

where $\mathcal{E}_{1,2}$ are given by expression (4), while the coefficients a_1 and a_2 may easily be found by means of formulae (5) and (6):

$$|a_1|^2 = \frac{I}{2}(1+A), \quad |a_2|^2 = \frac{I}{2}(1-A),$$

where $I = |a_1|^2 + |a_2|^2$ is the emission intensity in the optical fiber. Thus,

$$A = \frac{|a_1|^2 - |a_2|^2}{|a_1|^2 + |a_2|^2}.$$

Obviously parameter A may adopt values in the interval from -1 to +1. When the emission intensity is identical in both modes, $A = 0$. When exciting one of the eigenmodes when $A = \pm 1$ all the emission intensity is carried by a single eigenmode. Therefore parameter A may be called the excitation asymmetry parameter of the eigenmodes and the behavior of the polarization state of the quasi-monochromatic wave in the anisotropic single-mode optical fiber may be represented in the following manner. As emission propagates along a sufficiently extended optical fiber the mutual coherence of the optical fiber eigenmodes diminishes. This causes the periodic variations in the polarization state of the emission traveling along the optical fiber (present at the entrance to the optical fiber where the eigenmodes were coherent) to be replaced by a fixed polarization state on the axis of the optical fiber eigenmodes. In this case eigenmode emission is noncoherent, while the degree of polarization is determined by the asymmetry in eigenmode excitation, i.e., $P = |A|$. Therefore a necessary condition for complete depolarization of emission in a homogeneous anisotropic single-mode optical fiber is identical excitation of its eigenmodes.

Completely polarization emission is widely used in practice; the complex amplitude of such emission takes the form

$$\mathbf{E}(t) = R(t)\boldsymbol{\mathcal{E}}, \tag{39}$$

where $R(t)$ is a certain continuous scalar function of time; $\mathcal{E}$ is a fixed vector with complex components. Such a fixed-direction complex amplitude clearly determines the elliptically-polarized wave whose polarization ellipse varies in a like manner, i.e., without rotating or changing its shape. The degree of polarization of such emission is equal to unity, since $\det\,\rho = 0$, while the correlation tensor takes the form

$$\Gamma(\tau) = f(\tau)\rho, \tag{40}$$

where $f(\tau) = \langle R(t)R^*(t+\tau)\rangle / \langle |R(t)|^2\rangle$ is the correlation function of the emission.

We may easily determine that if arbitrary, partially-polarized emission is transmitted through a polarizing filter whose propagation matrix is equal to [9]

$$F_{\text{п}} = \begin{Vmatrix} \cos^2\theta & \frac{1}{2}\sin 2\theta \\ \frac{1}{2}\sin 2\theta & \sin^2\theta \end{Vmatrix},$$

where θ is the angle between the axis of the polarizing filter and one of the axes ox of oy, the correlation tensor will take the form of (40), since

$$F_{\text{п}}\Gamma(\tau)F_{\text{п}}^{+} = \left[\Gamma_{11}(\tau)\cos^2\theta + \Gamma_{22}(\tau)\sin^2\theta + \frac{\Gamma_{21}(\tau)+\Gamma_{12}(\tau)}{2}\sin 2\theta\right]F_{\text{п}}.$$

We should note that completely unpolarized emission will also have a correlation tensor of the type (40) in which $\rho_{12} = 0$, $\rho_{11} = \rho_{22}$. It is clear that for emission of the type (40) the correlation Stokes parameters take the form

$$\boldsymbol{\zeta}(\tau) = f(\tau)\boldsymbol{\xi}. \tag{41}$$

If the input emission takes the form (40) and its polarization state lies on the axis of the optical fiber eigenmodes, it is easily determined by means of (32), (33), and (41) that $B = 0$, $C = 0$. In this case the degree of polarization does not change along the optical fiber and is equal to $|A|$. In the case of completely polarized input emission of the type in (40) the degree of polarization at the exit will be

$$P = \{1 - [1 - f^2(\tau_1)](1 - A^2)\}^{1/2}. \tag{42}$$

Therefore emission that has been transmitted through a sufficiently long anisotropic single-mode optical fiber is partially depolarized if only the polarization state of the input emission does not lie on the axis of the optical fiber eigenmodes. The correlation tensor of emission that has passed through the optical fiber will no longer take the simple form of (40). Indeed, after calculating $\Gamma(\tau, z)$ from formulae (23) and (24) with satisfaction of conditions (26) and (27) we find that each component of the correlation tensor consists of the sum of terms $\sim f(\tau)$, $f(\tau + \tau_1)$ and $f(\tau - \tau_1)$, while the $\Gamma(\tau, z)$ tensor itself does not factor to (40). When condition (27) is violated and when it is necessary to incorporate the higher order polarization dispersion, the form of the correlation function also changes. We will consider this effect in greater detail and will find an expression for the correlation tensor components of emission that has been transmitted through the optical fiber.

As before we will take conditions (26) to be satisfied when the direction of the axis of the optical fiber eigenmodes is identical for all frequencies of the emission considered, although we will drop constraint (27). It is possible to separate conditions (26) and (27) since the first condition does not contain optical fiber length and the second may be violated for a sufficiently long anisotropic single-mode optical fiber. Assuming for simplicity that the correlation tensor of the input emission takes the form of (40), we obtain from formulae (23) and (24)

$$\Gamma_{11}(\tau, z) = \frac{f(\tau)}{2}(1 - \kappa A) + \frac{\varphi(\tau)}{4}\chi e^{2i\eta}M + \frac{\varphi^*(-\tau)}{4}\chi e^{-2i\eta}M^*, \tag{43}$$

$$\Gamma_{12}(\tau, z) = -\frac{f(\tau)}{2} i\chi e^{i\sigma}A + i\frac{\varphi(\tau)}{4} e^{2i\eta + i\sigma}(1 - \kappa)M - i\frac{\varphi^*(-\tau)}{4} \times$$
$$\times\, e^{-2i\eta + i\sigma}(1 + \kappa)M^*, \tag{44}$$

where

$$\varphi(\tau) = \int d\omega \exp[i\omega\tau + 2i\eta(\omega_0 + \omega) - 2i\eta(\omega_0)]\tilde{f}(\omega), \tag{45}$$

while $\tilde{f}(\omega)$ is the Fourier-transform of the input correlation function. The expressions for the remaining correlation tensor components may easily be found from formulae (43) and (44) by means of relations (16) and the equality

$$\zeta_4(\tau, z) \equiv \mathrm{Sp}\Gamma(\tau, z) = \mathrm{Sp}\Gamma(\tau, 0) = f(\tau), \tag{46}$$

which follows directly from expression (25).

We note that when exciting one of the optical fiber eigenmodes when $\xi(0) = \pm\mathbf{w}$ not only the polarization state but the entire correlation tensor Γ are conserved along the optical fiber. This follows directly from formulae (43), (44) accounting for the fact that $A = \pm 1$, $M = 0$.

In expression (45) we will set

$$2\eta(\omega_0 + \omega) - 2\eta(\omega_0) = \tau_1\omega + \tau_2^2\omega^2 + \tau_3^3\omega^3/3, \tag{47}$$

where $\tau_1 = d(2\eta(\omega_0))/d\omega_0$; $\tau_2^2 = d^2(\eta(\omega_0))/d\omega_0^2$; $\tau_3^3 = d^3(\eta(\omega_0))/d\omega_0^3$.

If we may neglect terms with higher order polarization dispersion τ_2 and τ_3 in expansion (47), then $\varphi(\tau) = f(\tau + \tau_1)$ and the correlation tensor components at the exit of the anisotropic single-mode optical fiber with an arbitrary emission polarization state at the entrance will consist of a superimposition of three functions, each of which is equal to the correlation function of the input emission centered for $\tau = 0$ and $\tau = \pm\tau_1$. When accounting for the higher order polarization dispersion the form of the function $\varphi(\tau)$ becomes other than $f(\tau)$, and the lateral peaks of the correlation tensor centered for $\tau = \pm\tau_1$ shift from the central peak which follows the form of the input correlation function.

In practice emission sources are encountered with a correlation function of the type

$$f(\tau) = \exp[-\tau^2(\Delta\omega)^2/4]. \tag{48}$$

for such emission at the entrance to the anisotropic single-mode optical fiber, when $|\tau_1\tau_c| \sim |\tau_2^2|$, although the third order polarization dispersion τ_3 is negligible, the lateral peaks of the correlation tensor at the exit take the form of functions of the type in (48), although with a greater width; by measuring this width it is possible to calculate τ_2. In the other case when $|\tau_2/\tau_c| \ll 1$ when we cannot neglect the third order polarization dispersion, we obtain from formula (45)

$$\varphi(\tau) = 2\sqrt{\pi}\,\nu^{-1}\exp(2\nu^{-6}/3)\exp[(\tau + \tau_1)\nu^{-2}/\tau_3]\,\mathrm{Ai}\left(\frac{\tau + \tau_1}{\tau_3} + \nu^{-4}\right), \tag{49}$$

where $\upsilon = |\tau_3\Delta\omega|$; $\mathrm{Ai}(x)$ is the Airy function. Here the lateral peaks of the correlation tensor consist of asymmetric nonmonotonic curves described by the Airy function. This case when τ_1 and τ_3 are important yet τ_2 is insignificant may be implemented, obviously, only at a certain frequency ω_0^*, where $\tau_2(\omega_0^*) = 0$. As the central frequency shifts from this value, condition $|\tau_2\Delta\omega| \ll 1$ is violated and the argument of the Airy function acquires the imaginary addition $2(\tau_2/\tau_3)^2\upsilon^{-2}$ which serves to "wash out" the minima of the Airy function.

Nonuniform optical fiber. In the preceding section the primary regularities in the variation of the polarization and coherence properties of emission traveling through an anisotropic single-mode optical fiber were derived after an exact solution of a coupled mode equation for a uniform optical fiber. In practice, however, anisotropic single-mode optical fibers whose parameters vary significantly along the optical fiber length are encountered. This is due to both fluctuations in their fabrication processes and due to the inhomogeneity of external influences. Since in the general case with β_1, β_2, h and σ dependent on z it is analytically difficult to solve the coupled mode equation, it is advisable to obtain equations for the observed quantities and to then carry out a general examination. Such observed quantities obviously include either the correlation tensor or the correlation Stokes parameters discussed in the previous section or their spectral densities. We will obtain equations for the latter. By multiplying the first equation from coupled mode equations (1) by $E_2^*(\omega', z)$, and by multiplying the equation complex-conjugate to the second equation for frequency ω' by $E_1(\omega, z)$ and summing, we find

$$\frac{d}{dz}[E_1(\omega,z)E_2^*(\omega',z)] = -i(\beta_1-\beta_2)[E_1(\omega,z)E_2^*(\omega',z)] - -h[E_1(\omega,z)E_1^*(\omega',z) - E_2(\omega,z)E_2^*(\omega',z)]. \quad (50)$$

Carrying out averaging over the ensemble of realizations of the random process and accounting for the fact that for the stationary emission $\langle E_i(\omega, z)E_j^*(\omega', z)\rangle = \tilde{I}_{ij}(\omega, \delta)\,\delta(\omega - \omega')$ and integrating both halves of the equation over ω' and accounting for the fact that the spectral density of the quasi-monochromatic emission $\tilde{J}_{ij}$ is related to I_{ij} by the relation $J_{ij}(\omega, z) = \tilde{I}_{ij}(\omega_0 + \omega, z)$, we obtain .

$$\frac{d\tilde{J}_{12}(\omega,z)}{dz} = -i(\beta_1-\beta_2)\tilde{J}_{12}(\omega,z) - h[\tilde{J}_{11}(\omega,z) - \tilde{J}_{22}(\omega,z)]. \quad (51)$$

Here β_1, β_2 and h are taken for frequency $(\omega_0 + \omega)$. The remaining three equations are derived analogously and these indicate that $\mathrm{Sp}J(\omega, z)$ is constant along the optical fiber, even if the optical fiber parameters vary along its length. Normalizing these equations to the total emission intensity $\mathrm{Sp}J(\tau = 0, z)$, we have

$$i\frac{d\Gamma(\omega,z)}{dz} = H(\omega',z)\Gamma(\omega,z) - \Gamma(\omega,z)H(\omega',z), \qquad \omega' = \omega_0 + \omega. \quad (52)$$

Here the Hermitian matrix $H(\omega_0 + \omega, z)$ is related to the matrix T figuring in equation (1) by the relation $H = iT$. It is obvious that equation (50) coincides with the quantum mechanical equation for the density matrix Γ of the system with Hamiltonian H in the basis of the two states. In this same basis equation (1) will obviously agree with the Schrodinger equation for the wave function E of the system with Hamiltonian H.

We obtain from equations (50) the following for the spectral density of the stokes vector $\zeta(\omega, z)$

$$d\zeta/dz = \Phi\zeta, \quad (53)$$

where the real matrix Φ takes the form

$$\Phi = \left\| \begin{matrix} 0 & -\Delta\beta & -2h\cos\sigma \\ \Delta\beta & 0 & 2h\sin\sigma \\ 2h\cos\sigma & -2h\sin\sigma & 0 \end{matrix} \right\|, \qquad (54)$$

$\Delta\beta = \beta_1 - \beta_2$. Analogous equations for monochromatic emission were obtained in study [10]. Due to the antisymmetry of Φ equations (51) may also be written differently:

$$d\zeta/dz = 2\gamma \mathbf{w} \times \zeta, \qquad (55)$$

Where the parameter $\gamma(\omega_0 + \omega)$ is determined in the preceding section; $\mathbf{w} = (-\chi\sin\sigma, -\chi\cos\sigma, \kappa)$ is the unit vector of the local eigenmode of the optical fiber. Obviously equation (55) describes the rotation of the vector $\zeta(\omega, z)$ in the Stokes space; the parameter 2γ determines the angular velocity of rotation while the vector $\mathbf{w}$ determines its direction.

For a uniform optical fiber the velocity of rotation is independent of z, although the different frequency components of $\zeta(\omega, z)$ rotate by different angles and therefore their superposition determining the Stokes vector ξ describes a spiral converging on the eigenmode axis (this is examined in the preceding section). If the direction of the axis of the optical fiber eigenmodes does not change along the optical fiber length, then by converting to the variable $t = 2\int_0^z dz' \gamma(z')$, in equation (53) we may again reduce the problem to a uniform optical fiber.

Clearly the vector ζ satisfying equation (53) may be situated continuously on the eigenmode axis, only if its direction does not change along the optical fiber.

Finally we will consider the situation where the direction of the eigenmode axis remains independent of frequency but changes significantly along the optical fiber. Here there is a continuous mismatch between the vector $\zeta(\omega, z)$ obeying equation (53) and the eigenmode axis, i.e., ζ precesses about the vector $\mathbf{w}$. If the optical fiber parameters vary lengthwise along the optical fiber such that uniform sections exist whose length is greater than the depolarization length $\tau_c/(2 d\gamma/d\omega_0)$, while the direction of the eigenmode axes of such sections are significantly different, the emission passing through such an anisotropic single-mode optical fiber will be depolarized independently of the polarization state of the input emission.

2. Experiment

The theoretical analysis carried out above was based on the concept of the correlation tensor: a quantity whose components are time-averages of the derivative of the Cartesian field components. This means that the physical quantities in the study are measured quantities and therefore the primary conclusions of the theory may easily be tested by experiment. An experimental investigation of the polarization and coherence properties of emission in an anisotropic single-mode optical fiber is of significant practical interest, since it makes it possible to establish the suitability of a specific anisotropic single-mode optical fiber in a particular communications link or device.

Experimental set-up. The experimental set-up is shown in Fig. 1. In order to expand the dynamic measurement range stimulated Raman scattering in an auxiliary optical fiber was used as the emission source. The pump source was an industrial Q-switched YAG LTN-101 laser (λ = 1.064 μm). Since in this case the stimulated Raman scattering is depolarized, it was possible by means of a monochromator and a Glan-Thompson prism to obtain linearly-polarized quasi-monochromatic emission with any azimuth in the 1.06-1.8 μm range. The coherence and polarization properties of such emission before and after the measured anisotropic single-mode optical fiber were analyzed by means of a modified Twyman-Green interferometer whose principle of operation and measurement technique were as follows. The emission exiting the optical fiber was split by the Glan-Thompson prism into two orthogonal components propagating independently in the two interferometer arms and after collection by a semitransparent mirror after the outcoupling Glan-Thompson prism (analyzer) they produced a fringe pattern. In order to expand the dynamic measurement range the interferometer was tuned to an infinitely broad band and all emission transmitted through the analyzer was collected by a photodetector. An optical delay prism was placed at a piezoceramic element to which a sawtooth voltage was injected which made it possible to modulate the length of one of the interferometer arms within a few wavelengths. Due to this modulation of the length of the interferometer arm, the interference pattern "ran" across the detector and the modulation level of the signal from

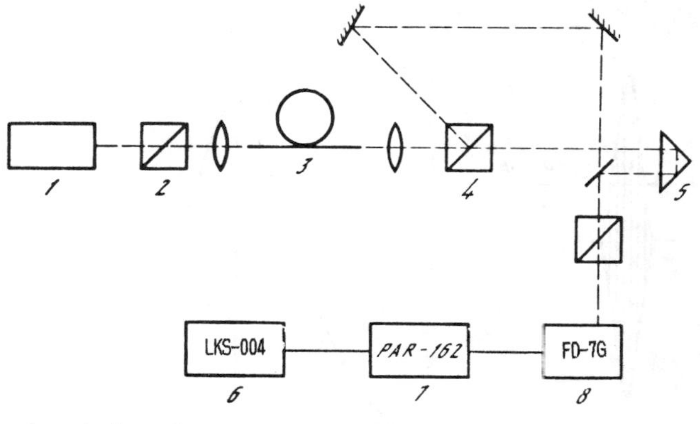

Fig. 1. Experimental set-up.
1 - source; 2 - polarizer; 3 - test single-mode optical fiber; 4- Glan-Thompson prism; 5 - optical delay prism; 6 - printer; 7 - synchronous amplifier; 8 - photodetector

the detector was proportional to the visibility υ of the interference pattern. By varying the length of one of the arms by displacement of the optical delay prism the $\upsilon(\tau)$ relation was determined (if d is the difference of the interferometer arms, then $\tau = 2d/c$), which is equal to the modulus of the correlation tensor of the emission [11].

Initially the measurements were carried out on a short optical fiber section in order to calibrate the interferometer, i.e., to find the prism position in which the interferometer arms were equal. This measurement also made it possible to determine the emission correlation function of the source. Fig. 2a gives a standard pattern obtained on the printer in such a case. The difference of the arm lengths d was measured in discrete steps of 10 μm each (τ = 0.067 ps). The maximum modulation level obviously corresponds to an identical length of the interferometer arms (τ = 0). The envelope of the derived pattern is the emission correlation function. As we know [12] in the case of a stationary random process the emission correlation function and the spectrum of the source are related by a Fourier transform. Fig. 2b gives the emission spectrum whose correlation function is shown in Fig. 2a. A numerical Fourier transform of the correlation function yielded a result that agrees with the spectrum within 5% shown in Fig. 2b. This confirms the measurement accuracy.

The measurements were then carried out on long optical fiber sections in which the emission correlation function had, as a rule, a number of peaks with the maximum peak at τ = 0 representing the emission correlation function of the source while the remaining peaks represented the product of the interaction between the "fast" mode of the optical fiber with the "slow" mode, i.e., the mutual coherence function $\varphi(\tau)$ and bore information on the polarization characteristics of the emission. The accuracy in determining the temporal position of the mutual coherence function was better than 0.1 ps, i.e., the first order polarization dispersion was measured with such an error level.

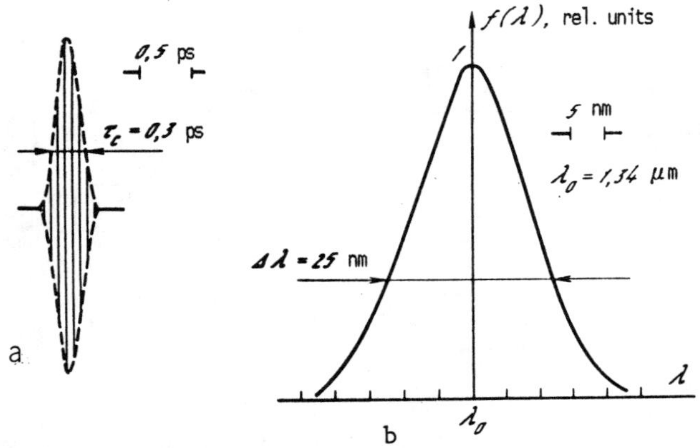

Fig. 2. The correlation function (a) and spectrum (b) of the emission source

It should be noted that it was necessary to carry out measurements of the emission correlation function in the short section only once to calibrate the interferometer. In subsequent measurements the peak corresponding to the emission correlation function of the source could easily be identified since the length of the variable arm of the interferometer corresponding to $\tau = 0$ was known.

The correlation function of emission propagating along an anisotropic single-mode high birefringence optical fiber. Fig. 3 shows the emission correlation function measured in a V28-E1 optical fiber (z = 120 m, $\Delta n = 6.5 \cdot 10^{-3}$, $2a = 4$ μm) at 1.24 μm. As indicated by the theoretical analysis the distance between the lateral and central peaks of the emission correlation function corresponds to the first order polarization dispersion time τ_1.

It is clear from the relation that the lateral peak is broader than the central peak (the central peak is the correlation function of the emission source), and the width of this peak is dependent on the wavelength [13]. As established from the theoretical analysis the second order polarization dispersion τ_2 broadens the lateral peak and therefore it is possible to directly determine the value of τ_2 from emission correlation function measurements. Since the spectrum of the emission source is Gaussian, we obtain a simple relation between the coherence time of the source τ_c and the characteristic time (at e^{-1}) of the emission correlation function τ_b and the second order polarization dispersion τ_2. Using expression (45) and bearing in mind that the amplitude of the mutual coherence function is proportional to $|\varphi(\tau)|$, we have

$$\tau_2^2 = \frac{\tau_c}{8}\sqrt{\tau_b^2 - \tau_c^2}. \tag{56}$$

The magnitude of the broadening of the lateral emission correlation function peak depends on the spectral width of the emission and the optical fiber length and therefore Fig. 4 gives the normalized τ_2 relation in units of ps/(nm·km).

In this case the analytic expression for the mutual coherence function takes the form

$$\varphi(\tau) = A \exp\left[-\frac{(\tau + \tau_1)^2}{\tau_c^2/4 + 16\tau_2^4/\tau_c^2}\right]. \tag{57}$$

It is clear from this expression that there exists a value of τ_c where τ_b is at a minimum, specifically:

$$\tau_c^{opt} = 2\sqrt{2}\,\tau_2, \quad \tau_b^{min} = \sqrt{2}\,\tau_c^{opt}$$

Physically this may be explained in the following manner. The characteristic time of the mutual coherence function at the exit of the anisotropic single-mode optical fiber is determined by the emission correlation function of the source and the frequency properties of the optical fiber. The shorter the coherence time (the narrower the emission correlation function), the broader the spectrum and the greater the dispersion of the relative delay of the optical fiber eigenmodes; on the other hand, an increase in τ_c will serve to broaden the emission correlation function and therefore the mutual coherence function. Therefore there will exist an optimum coherence time where the influence of both processes is at a minimum. Naturally τ_c^{opt} depends on τ_2. An illustration of this process may be found in Fig. 5 which shows the $\tau_b(\tau_c)$ relation (V28-E2 optical fiber, λ = 1.5 μm).

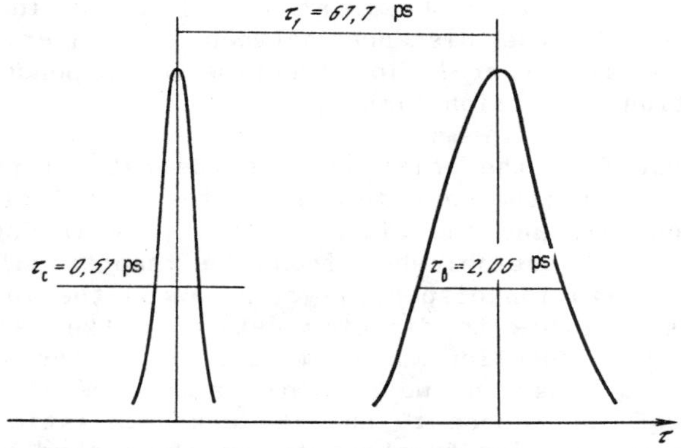

Fig. 3. The correlation emission function at the exit of a high birefringence optical fiber

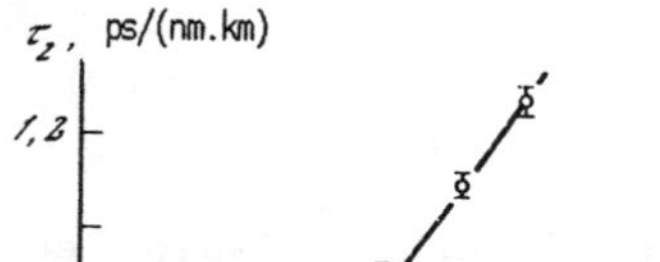

Fig. 4. The spectral dependence of the second order polarization dispersion in the V28-E1 optical fiber

Fig. 5. The characteristic time of the mutual coherence function plotted as a function of the coherence time of the source

The emission correlation function in single-mode low birefringence optical fibers. Fig. 6 gives measurement results for the emission correlation function in a GDD-71 optical fiber (z = 1050 m, Δn =

10^{-2}, $2a$ = 6 μm) at a number of wavelengths. As indicated from the results the emission correlation function is significantly more complex than the emission correlation function in high birefringence optical fibers (see Fig. 3). At a number of wavelengths the emission correlation function manifested a clearly-expressed oscillatory nature, while the peak of the curve did not coincide with zero delay between the waves.

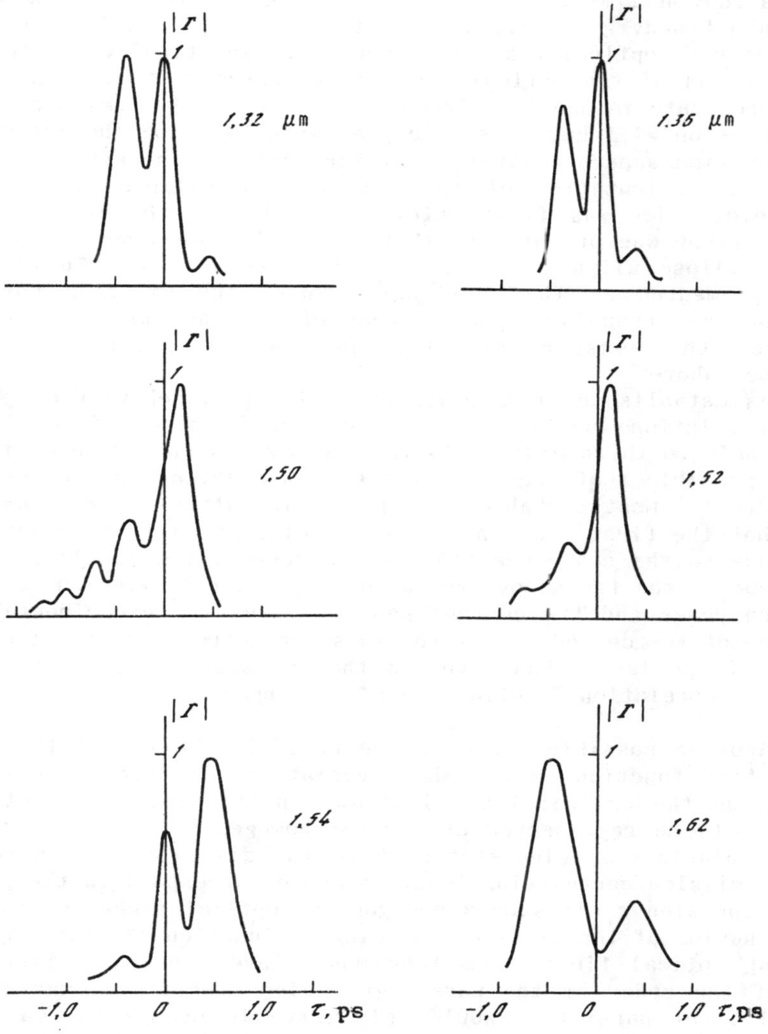

Fig. 6. Correlation function of emission at the exit of a low birefringence optical fiber

Here we should say a few words regarding the measurement technique. Overall it was the same as that described above employed to investigate the emission correlation function in a high birefringence anisotropic single-mode optical fiber, although due to the low τ_1 the measurement technique in this case had a number of special features which may be described as follows. Clearly if τ_1 is comparable to τ_c then if all three peaks are present in the emission correlation function (the central autocorrelation peak and two lateral peaks corresponding to cross-correlation), it is rather difficult to recover information on τ_1m, since the emission correlation function will appear as a broad, generally asymmetric peak [13]. An ideal case would be where both optical fiber eigenmodes are identically excited, while the exit end of the optical fiber is oriented so that each eigenmode would propagate in one interferometer arm. Then the emission correlation function will have a single peak which will be the mutual coherence function separated by τ_1 from zero delay. Therefore in the test sample the orientation of the polarization ellipse of the optical fiber eigenmodes was first determined and then the exit end of the optical fiber was oriented so that the semi-major axes of the polarization ellipse aligned with the interferometer axes. This procedure was implemented at the wavelengths where the emission correlation function was irregular. In the case of a clear emission correlation function the measurement technique was no different from that described above.

As established previously the oscillatory behavior of the emission correlation function in the single-mode optical fiber is caused by the nonzero third order polarization dispersion. However it is entirely possible that the oscillations could arise from another cause. There are two most probable causes of this nature. First, due to the fact that the first order polarization dispersion τ_1 is comparable in magnitude to the coherence time of the source, the oscillations could arise due to the frequency overlap of the lateral emission correlation function peaks and the central peak. However an experimental investigation of the dependence of the emission correlation function on the optical fiber length has revealed that no such oscillations occurred when the correlation function peaks overlapped.

Another possible cause of the unusual behavior of the emission correlation function is the sharp variation in optical fiber parameters along the optical fiber length. In this case the test optical fiber could be represented as several homogeneous segments with different polarization eigenstates while the distance between the peaks of the emission correlation function would correspond to the polarization dispersion τ_1 in such a homogeneous optical fiber section. With such behavior of the emission correlation function the inhomogeneities altering optical fiber parameters would have a characteristic dimension of the order of the wavelength, since continuous variations in optical fiber parameters would only serve to change the value of τ_1 compared to the case of a homogeneous optical fiber and would not create additional peaks. On the other hand, such inhomogeneities would cause a sharp increase in optical fiber losses and would be somewhat independent. The test sample losses were less than 2 dB/km in the 1.1-1.6 μm range (excluding the absorption peak at 1.39 μm), while the form of the emission correlation function was dependent on the wavelength. Therefore we may conclude that the cause of the oscillations is the third order polarization dispersion. This is also indicated by measurements of the emission correlation function along two different optical fiber sections; in two measurements of the sample the emission correlation function was identical while the distance between the peaks dropped by approximately 10%. As demonstrated theoretically in the case $\tau_2 = 0$, $\tau_3 \neq 0$ the mutual coherence function is described by an expression of the type

$$\varphi(\tau) = a \,\mathrm{Ai}\left[\frac{\tau + \tau_1}{\tau_3} + \left(\frac{\tau_c}{4\tau_3}\right)^4\right].$$

Since $\tau_3 \simeq \tau_c$ and $\tau_3 \sim z^{1/3}$, an increase in optical fiber length by a factor of n would serve to increase the distance between the maxima of the Airy function by a factor of $n^{1/3}$ which was clear in experiment.

Similar behavior of the emission correlation function was observed in all anisotropic single-mode low birefringence optical fiber samples, although for each sample the wavelength and azimuth of the input linear-polarized emission were individual quantities determined both by the spectral dependence of τ_1 and the orientation of the polarization ellipse of the eigenmodes of the optical fiber with respect to the interferometer axes. Thus, for example, in the GDD-71 optical fiber the spectral dependence of the first order polarization dispersion appeared as shown in Fig. 7 [13]. Since in the 1.5 μm range the second order polarization dispersion τ_2 is equal to zero while τ_3 is close to its maximum, the mutual coherence function at this wavelength will have a clearly-expressed oscillatory nature, which is confirmed by the experimental results shown in Fig. 6.

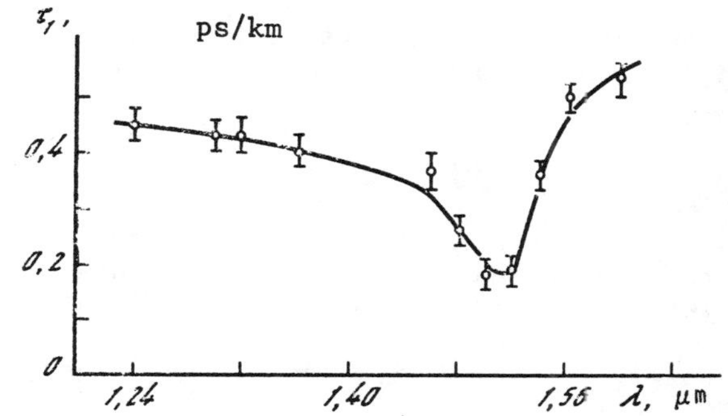

Fig. 7. The spectral dependence of the first order polarization dispersion in a GDD-71 optical fiber

It is interesting to note that the behavior of the mutual coherence function in single-mode optical fibers is largely similar to the behavior of a short optical pulse in the optical fibers [14]. This is not an arbitrary occurrence, as is clear in many studies (see, for example, [11]): the mutual coherence function obeys the wave equation which is also obeyed by the optical pulse envelope. In other words in addition to electromagnetic waves, correlation waves are also propagating in the medium. Like the influence of the medium on the electromagnetic wave, the correlation waves are subject to an analogous action.

Conclusion

The primary results from the theoretical and experimental investigation of the coherence and polarization properties of quasi-monochromatic emission in single-mode optical fibers may be generalized in the following manner.

1. A model of a regular dual-mode optical fiber was used to carry out a theoretical analysis of the coherence and polarization properties of quasi-monochromatic emission propagating along an anisotropic single-mode optical fiber. Expressions were derived relating the polarization characteristics of the emission (the Stokes parameters) at the exit and entrance to the optical fiber to the coherence properties of the emission source. The influence of the polarization properties of the optical fiber on the mutual coherence function was investigated and it is demonstrated that a single-mode optical fiber is a depolarizing optical system.

2. A technique is developed for measuring the correlation function of emission in single-mode optical fibers over a broad spectral range. The spectral dependence of the correlation function of emission in large and small birefringence samples was investigated. The spectral dependence of the second order polarization dispersion was measured and the third order polarization dispersion was estimated.

3. All experimental results were satisfactorily explained by the theory based on the regular single-mode optical fiber model.

Bibliography

1. Snyder, A.W. "Asymptotic expression for eigenmodes and eigenvalues of a dielectric or optical waveguide" IEEE TRANS. MICROWAVE THEORY AND TECH., 1969, V. MTT-77, No. 12, p. 1230-1138.

2. Gloge, D. "Weakly guiding fibers" APPL. POT., 1971, V. 10, No. 10, p. 2252-2258.

3. Marcuse, D. "Theory of dielectric waveguides" NY: Acad. Press, 1974, 435 p.

4. Snyder, A.W., Love, D. "Optical waveguide theory" NY: Acad. Press, 1983, 734 p.

5. Sakai, J.I., Kimura, T. "Birefringence and polarization characteristics of single-mode optical fiber under elastic deformation" IEEE TRANS. QUANT. ELECTR., 1981, V. 17, No. 6, p. 1041-1051.

6. Azzam, R., Bashara, N. "Ellipsometriya i polyarizovannyy svet" [Ellipsometry and polarized light] Moscow: Mir, 1981, 583 p.

7. Landau, L.D., Lifshits, E.M. "Teoriya polya" [Field theory]y Moscow: Nauka, 1967, p. 460.

8. Perina, Ya. "Kogerentnost' sveta" [The coherence of light] Moscow: Mir, 1974, 386 p.

9. Barakat, R. "Theory of the coherency matrix for light of arbitrary bandwidth" J. OPT. SOC. AMER. 1963, V. 53, No. 3, p. 317-323.

10. Kozel, S.M., Listvin, V.N., Shatalin, S.V. "Investigation of the polarization anisotropy of single-mode silica fiber" IZV. VUZOV. RADIOFIZIKA, 1984, V. 27, No. 4, p. 505-511.

11. Born, M., Vol'f, E. "Osnovy optiki" [The principles of optics] Moscow: Nauka, 1973, 720 p.

12. Rytov, S.M. "Vvedenie v statisticheskuyu radiofiziku. T. q. Sluchaynye protsessy" [An introduction to statistical radiophysics] V. 1, Random processes], Moscow: Nauka, 1976, 494 p.

13. Grudinin, A.B. "The coherence and polarization properties of radiators in single-mode optical fibers": Dissertation for the Candidates Degree in Physics and Mathematics, 01.04.04 - Moscow-:IOFAN, 1985.

14. Marcuse, D. "Pulse distortion on single-mode fibers" APPL. OPT. 1980, V. 19, No. 10, p. 1653-1660.

Problems of Data Transmission by Optical Solitons

A.S. Belanov, Ye.A. Golovchenko, Ye.M. Dianov, Z.S. Nikonova, A.M. Prokhorov, V.N. Serkin

Abstract: This study considers the primary physical operating principles of "nonlinear" fiber-optic communications links compared to the maximum capabilities of "linear" systems. It is demonstrated that the central problem in data transmission by solitons on optical fibers is the problem of minimizing optical fiber loss. The possibility of achieving soliton states in the medium IR in optical fibers is discussed as well as the problem of nonlinear soliton interaction in the transmission line; the pair interactions between solitons in linear loss optical fibers are examined. A method of stimulated Raman amplification of solitons is proposed and developed theoretically. It is demonstrated that in the transition to the femtosecond range of pulse durations the fundamental role is played by effects associated with the nonstationarity of the nonlinear field-cubic polarization of the medium. It is established that stationary isolated pulses may arise in the optical fiber with a duration of the order of a few optical periods.

Introduction

The development of single-mode optical fibers with vanishingly-low losses has shifted the accent from optical losses as the limiting factor in the development of high-capacity fiber-optic communications links to the bandwidth of the optical fiber.

Fiber-optic communications is characterized by a higher transmitted data density in both space and time. A broadband parameter of ~240 GHz·km [1] has been achieved in data transmission on a single-mode optical fiber. Although this is a very high figure it only slightly reflects the potential broadband parameter of existing optical fibers. Indeed, the spectral bandwidth of currently-manufactured optical fibers lies between 0.8 to 1.8 μm corresponding to $\Delta\upsilon = 2\cdot10^5$ GHz which is nearly three orders of magnitude greater than the bandwidth determined by the broadband parameter.

Intermodal, material and waveguide dispersion all hinder the expansion of the broadband parameter of a single-mode optical fiber. Intermodal dispersion results from the difference in the delay of two orthogonally-polarized propagating modes corresponding to the fundamental mode. In the case of an ideal uniform isotropic optical fiber of circular cross-section there is no dispersion between these modes. The dispersion will appear when, for example, the optical fiber core becomes elliptical. Single-mode optical fibers that conserve polarization make it possible to eliminate intermodal dispersion.

Material dispersion may be attributed to a change in the refractive index of the optical fiber material with frequency. For silica-based optical fibers this factor is minimized at $\lambda = 1.3\ \mu m$.

The waveguide or structural dispersion of an optical fiber is manifest even when there is no wavelength dependence of the refractive index and is attributable to the influence of the geometric structure of the optical fiber. For example, it is dependent on the core diameter for a circular optical fiber.

Today it is possible to identify two primary approaches to solving the problem of minimizing dispersion broadening of pulses in optical fibers. The first technique involves using broadband optical fibers with compensated first and second order dispersion [2-11].

Minimizing the first and second order dispersion coefficients by appropriate selection of the primary parameters of the optical fibers makes it possible not only to increase the broadband parameter of the optical fiber but also to fabricate optical fibers in which the "zero" dispersion range coincides with minimum optical losses. It should be emphasized that the principal difficulties in this method of minimizing dispersion broadening of pulses in fiber-optic communications links are related both to insufficiently-developed optical fiber technology and the requirement for highly-precise tuning of the laser emission wavelength to the "zero" dispersion point of the group velocity. As indicated by results from study [5] a possible 1% "drift" in the wavelength of the master laser from its optimum value reduces the data transmission speed to tens of gigabits per second over an optical fiber length on the order of 10 km.

A new possibility for solving the problem of "zero dispersion" pulse propagation in fiber-optic communications links is based on using the nonlinear effects in the optical fiber and implementing so-called soliton states of pulse propagation in optical fibers [12-44].

The uniqueness of the soliton deriving from the fact that the dispersion pulse broadening is precisely equalized by its nonlinear self-steepening allowed the authors of studies [12, 13] to propose the optical soliton as the information medium in fiber-optic communications systems. The corresponding fiber-optic communications systems should therefore be called nonlinear systems. Even the initial calculations carried out in study [13] revealed that in such "nonlinear" communications lines the data transmission speed could be an order of magnitude greater than the best "linear" fiber-optic communications links operating with "zero" group velocity dispersion and could achieve a terabit of data per second over a distance of 30 km. Interest in the problems and prospects for the development of nonlinear optical soliton communications system was also stimulated by the fact that as picosecond laser technology developed, optical solitons became an accessible source of experimental research. Therefore in the present study we will focus primarily on the fundamental physical principles of data transmission by optical solitons and the unresolved problems that remain the key issues in developing "nonlinear" fiber-optic soliton communications links.

This study examines the fundamental physical operational principles of "nonlinear" fiber-optic communications links compared to the maximum capacities of "linear" systems. It is demonstrated that the central problem in information transmission by solitons on optical fibers is minimizing optical losses in the optical fiber. Unlike "linear" communications systems in which the data transmission speed is limited by the group velocity dispersion, in soliton pulse propagation the rate of data transmission is limited by losses in the optical fiber and the nonlinear interaction of solitons over the pulse period. The possibility of achieving soliton states in the medium IR in optical fibers is discussed. The problem of minimizing the power required for soliton generation in fiber-optic communications links is discussed and the primary physical properties of data transmission by higher order solitons are examined. The problem of nonlinear soliton interaction in the transmission link is discussed and pair soliton interactions in linear loss optical fibers are considered. The dynamics of soliton generation in optical fibers is investigated and the critical role of amplitude-phase perturbations and optical losses in the formation of coupled soliton states in optical fibers is demonstrated. A method is proposed and developed theoretically for soliton gain by the stimulated Raman scattering technique. It is demonstrated that in the transition to the femtosecond range of launched pulses the fundamental role belongs to effects associated with the non-stationarity of the nonlinear field-cubic polarization of the medium.

It is established that stationary, remote pulses with a duration in the order of a few periods of the light waves may exist in optical fibers; this indicates the possibility of achieving data transmission speeds by optical solitons up through the critical level determined by the spectral bandwidth of the optical fiber.

1. Maximum Data Transmission Speed in the Case of Linear Propagation of Optical Pulses Through an optical Fiber

The uniqueness of an optical soliton as a unit bearing information in a fiber-optic cable is most clearly manifest in a comparison to traditional "linear" data transmission techniques. We will examine in greater detail the primary approaches to solving the dispersion broadening of pulses in field linear propagation through an optical fiber.

Using the familiar theory describing pulse propagation in optical fibers [3-5] we may demonstrate that a Gaussian pulse with an envelope $E(t) \sim \exp(-2t^2/\tau_0^2)$($\tau_0$ is the input pulse duration measured at e^{-1} of its maximum power), with a spectral linewidth of $\Delta\lambda_и$ and a pulse linewidth of $\Delta\lambda_0 = 2\lambda^2/\pi c\tau_0$ will after passing through an optical fiber of length L have a duration of

$$\tau = [\tau_0^2 + (\Delta\tau_1)^2 + (\Delta\tau_2)^2 + (\Delta\tau_3)^2]^{1/2}, \tag{1}$$

where

$$\Delta\tau_1 = S_1 L\Delta\lambda; \quad \Delta\tau_2 = (2S_1 + S_2\lambda)L\Delta\lambda^2/(4\lambda); \tag{2}$$

$$\Delta\tau_3 = (6S_1 + 6S_2\lambda + S_3\lambda^2)L(\Delta\lambda)^3/(8\lambda^2); \tag{3}$$

$$\Delta\lambda = [(\Delta\lambda_0)^2 + (\Delta\lambda_и)^2]^{1/2}; \tag{4}$$

$$S_1 = c^{-1}\frac{dn_{гр}}{d\lambda}; \quad S_2 = c^{-1}\frac{d^2 n_{гр}}{d\lambda^2}; \quad S_3 = c^{-1}\frac{d^3 n_{гр}}{d\lambda^3} \tag{5}$$

are the first, second, and third order dispersion coefficients;

$$n_{гр} = c/\upsilon_{гр} = n_э - \lambda dn_э/d\lambda; \quad n_э = c/\upsilon_ф;$$

c is the speed of light in the vacuum: $\upsilon_{гр}$, $\upsilon_ф$ are the group and phase velocities of the mode in the optical fiber.

The duration of pulses τ_0 employed in optical communications links is almost always greater than 1 ps and therefore with a sufficiently narrow spectral linewidth of the source emission the condition $\Delta\lambda \ll \lambda$ is always satisfied. In this regard it follows from (3), (4) that in the general case $|\Delta\tau_1| > |\Delta\tau_2| > |\Delta\tau_3|$. However there exist ranges of values of the primary parameters of the optical fiber in which it is possible to achieve $\Delta\tau_1$ and even $\Delta\tau_1 = \Delta\tau_2 = 0$. In the latter case pulse distortion will be minimized and the pulse duration product of the optical fiber will be maximized. We will consider the possibility of minimizing dispersion pulse broadening for the case of linear propagation through an optical fiber in greater detail.

We will first consider the case where $S_1 \neq 0$ and, consequently, $\Delta\tau_1 \neq 0$. Neglecting $\Delta\tau_2$ and $\Delta\tau_3$ in (1), we find that

$$\tau = [\tau_0^2 + (S_1\Delta\lambda L)^2]^{1/2}. \tag{6}$$

An analysis of (6) reveals that there exists an optimum input pulse duration of

$$\tau_{opt\,1} = (2\lambda^2 S_1 L/\pi c)^{1/2},$$

at which the duration of the output pulse will be minimized:

$$\tau = [2\tau^2_{opt1} + (S_1 L \Delta\lambda_н)^2]^{1/2}.$$

Specifically when $\Delta\lambda_H = 0$ we obtain the familiar result [3]

$$\tau = \sqrt{2}\tau_{opt1} = (4\lambda^2 S_1 L/\pi c)^{1/2}; \tag{7}$$

i.e., the duration of the exit pulse and the rate of data transmission are proportional to $\sqrt{L}$.

We will consider the second case when it is possible to obtain S_1 and, consequently, $\Delta\tau_1 = 0$. According to (2) $\Delta\tau_1 \sim S_1$. For small differentials in the refractive indices of the core and the cladding $\Delta n = n_1 - n_2$ in accordance with study [6]

$$S_1 = -\frac{1}{\lambda c}\left(\lambda^2 \frac{d^2 n_{rp}}{d\lambda^2} + \Delta n V \frac{d^2 VB}{dV^2}\right), \tag{8}$$

where $B = (n_3^2 - n_2^2)/(n_1^2 - n_2^2) \simeq (n_3 - n_2)/(n_1 - n_2)$; $V = 2\pi a \lambda^{-1}\sqrt{n_1^2 - n_2^2}$.

For the given wavelength λ and the selected optical fiber material we may find such values of V, $\Delta n = n_1 - n_2$, $2a$, $2b$ where consistent with (8) the influence of material dispersion will be compensated by the waveguide dispersion of opposite sign. Pulse distortions in this case will be minimized.

Indeed, taking $\Delta\tau_1 = 0$ in (1) and neglecting $\Delta\tau_3$, we find that

$$\tau = \{\tau_0^2 + [S_2 L (\Delta\lambda)^2/4]^2\}^{1/2}.$$

An analysis of this formula reveals that the exit pulse duration will be minimized with an optimum duration of the entrance pulse [7]:

$$\tau_{opt2} = (\lambda^4 S_2 L/\pi^2 c^2)^{1/3}[(1 + \sqrt{1 + A})^{1/3} + (1 - \sqrt{1 + A})^{1/3}]^{1/2},$$

where

$$A = -(\pi c S_2 L/\lambda^2)^2[(\Delta\lambda_н)^2/6]^3.$$

Where $\Delta\lambda_H \to 0$m, then

$$\tau_{opt2} = (\sqrt{2}\lambda^4 S_2 L/\pi^2 c^2)^{1/3}$$

and the duration of the exit pulse will be

$$\tau = (1{,}5\,\tau_{opt2})^{1/2} = (1{,}5\sqrt{3}\lambda^4 S_2 L/\pi^2 c^2)^{1/3},$$

i.e., proportional to $\sqrt[3]{L}$. In this case a data transmission speed of $M = (\tau \; 22|\ln\delta|)^{-1/2}$ (δ is the relative intensity overlap of sequential pulses) of several tens of gigabits per second over a distance of 100 km is possible.

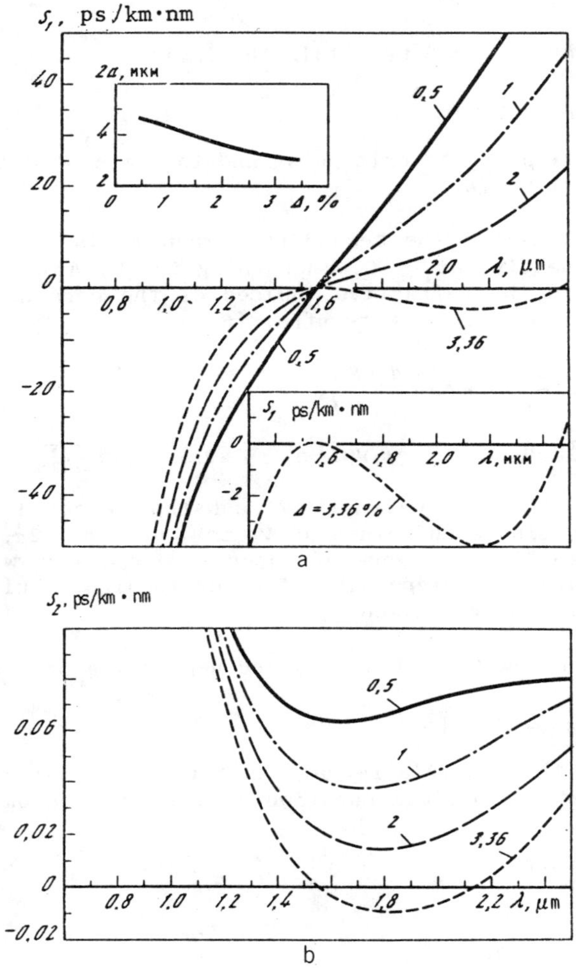

Fig. 1. The first and second order dispersion coefficients plotted as a function of wavelength for dual-layer optical fibers
The numbers labeling the curves represent the values of Δ in %.

Fig. 1 gives calculation results for s_1 (Fig. 1a) and S_2 (Fig. 1b) for dual-layer step refractive index optical fibers for different values of $\Delta = (n_1 - n_2)/n_2$ when $S_1 = 0$ for $\lambda = 1.55$ μm. It was assumed that the cladding was fabricated from silica doped by three mol-% B_2O_3. It is clear from the graph that with growth of Δ the broadband parameter increases and reaches a maximum value at Δ = 3.36% and Δ' = 3.2 μm when the coefficient S_1 does not exceed ±5 ps/km·nm) in the range λ = 1.3-2.6 μm (V = 2.89-1.43). This same diagram gives the core diameter of the optical fiber $2a$ plotted as a function of Δ. It is clear from an analysis of the graph that in order to develop broadband optical fibers it is necessary to fabricate optical fibers with very low core diameters and small differences in the refractive indices of the core and the cladding. The maximum broadband parameter is achieved when the second order dispersion coefficient $S_2 = 0$.

For Δ = 3.36% and λ = 1.55 μm $S_2 = 0$. In this case pulse distortion is entirely dependent on the third order dispersion coefficient. Indeed it follows from (1)-(5) that when $S_1 = 0$ and $S_2 = 0$ the duration of the exit pulse is

$$\tau = [\tau_0^2 + (S_3 L/8)^2 (\Delta\lambda)^2]^{1/2},$$

and it is minimized with an optimum exit pulse duration of

$$\tau_{\mathrm{opt}\,3} = (\sqrt{3} S_3 L \lambda^2 (8\pi c)^{-1} \{ (\Delta\lambda_{\mathrm{H}})^2 + [(\Delta\lambda_{\mathrm{H}})^4 + 64\lambda^2 (\sqrt{3} S_3 L \pi c)^{-1}]^{-1/2} \})^{-1/2}.$$

Specifically, when $\Delta\lambda_{\mathrm{H}} = 0$

$$\tau_{\mathrm{opt}\,3} = [\sqrt{3} S_3 L (\lambda^3/\pi c)^3]^{1/4}, \quad \tau = 2\tau_{\mathrm{opt}\,3}/\sqrt{3}, \tag{9}$$

i.e., the duration of the exit pulse is proportional to $\sqrt[4]{L}$. In this case a transmission speed of several hundred gigabits per second is possible.

The most suitable broadband optical fibers are three-layer optical fibers, particularly W-type optical fibers and optical fibers of circular cross-section.

We should note that when using sources with a broad spectral linewidth $\Delta\lambda_{\mathrm{H}} \gg \Delta\lambda_0$ the duration of the exit pulse $S_1 \Delta\lambda_{\mathrm{H}} L$, $S_1 = 0$; $\tau_2 = 0.25 S_2 (\Delta\lambda_{\mathrm{И}})^2 L$, $\Delta S_1 = 0$, $s_2 \neq 0$; $\tau_3 = 0.125 S_3 (\Delta\lambda_{\mathrm{И}})^3 L$, $S_1 = S_2 = 0$, $S_3 \neq 0$, i.e., it is independent of the duration of the entrance pulse τ_0 and is proportional to the optical fiber length L. Calculations revealed that for L = 10 km $\Delta\lambda_{\mathrm{И}}$ = 40 nm the following exit pulse durations will be obtained: τ_1 = 1-20 ns with $S_1 = 0$, τ_2 = 20-200 ps, $S_1 = 0$, $s_2 \neq 0$.

Narrowband sources to provide optimum entrance pulse duration and compensated dispersion optical fibers make it possible to achieve maximum possible data transmission speeds of hundreds of gigabits per second over a distance of 100 km. These figures represent the maximum levels for field-linear pulse propagation in an optical fiber. It is possible to further increase data transmission speed in an optical fiber in nonlinear pulse propagation by using so-called soliton effects in optical fibers.

2. The Physical Picture of Soliton Formation in a Single-Mode Optical Fiber

The concept of solitons which is a central concept in modern physics has had a rather long history of development for application to fiber-optics. Single-mode silican optical fibers were proposed for optical soliton experiments as early as study [12]. However almost a decade was required before physical experiments involving optical solitons became a reality [14]. This was due to the technological difficulties of fabricating single-mode optical fibers with vanishingly low losses in the 1.3-1.5 μm spectral range corresponding to negative ("anomalous") group velocity dispersion as well as problems of fabricating stable picosecond light sources that are tunable in this spectral range.

We will examine the primary physical effects that occur from the propagation of powerful picosecond laser pulses in optical fibers. In this case we will limit our examination to a simple model. We will represent the optical fiber in the form of a series of sufficiently thin layers ("disks") with purely nonlinear and dispersive properties and will consider what occurs with a pulse launched into the optical fiber as it passes sequentially through a series of "nonlinear" and "dispersion" layers of thickness $\Delta z_{\text{НЛ}}$, $\Delta z_{\text{Д}}$. We will assume that the phase continuity conditions of the lightwave are satisfied as it passes through each of the "disks" of thickness Δz.

Such a breakdown of the problem physically into layers with purely nonlinear and dispersive properties not only yields a simple and convenient picture of the nonlinear propagation of pulses in an optical fiber but also makes it possible to formulate effective numerical algorithms for solving the corresponding problems on a computer.

We will introduce a quantitative measure of the nonlinearity of the refractive index of the optical fiber material (including the outer cladding):

$$n = n_0 + \tfrac{1}{2}\, n_{2\,\text{эф}} |E|^2 .$$

In glasses the nonlinear addition to the refractive index is due to the electron Kerr effect and occurs over a time period in the order of the Bohr orbit ~10^{-15} s. Unlike an infinite uniform bulk medium where the nonlinear susceptibility is constant in the beam cross-section, the effective susceptibility in an optical fiber is dependent on the transverse distribution of the mode launched in the fiber-optic cable and is determined by the average value in the optical fiber cross-section, i.e.,

$$n_{2\,\text{эф}} = \frac{\int_0^\infty n_2 R^4(r) 2\pi r dr}{\int_0^\infty R^2(r) 2\pi r dr} \equiv a n_2 .$$

here $R(r)$ is a function describing the spatial field distribution in a linear approximation: α is the geometric factor; n_2 is the nonlinear addition to the refractive index of the homogeneous phase. Therefore the nonlinearity of $n_2|E|^2$ depends on the radial field distribution in the optical fiber cross-section.

In the case of a dual-layer optical fiber the fundamental LP_{01} mode is a cylindrically-symmetric wave described by a Bessel function within the optical fiber core and by a Macdonald function in the cladding:

$$R(r) = A_0 J_0(UR), \quad R < 1,$$
$$R(r) = A_0 J_0(U) K_0(WR)/K_0(W), \quad R > 1,$$

where $R = r/a$ is the normalized variable, while the eigenvalues U and W are determined from the relations

$$U^2 = a^2(n_1^2 K_0^2 - K^2), \quad W^2 = a^2(K^2 - n_2 K_0^2).$$

Here

$$V^2 = U^2 + W^2 = a^2 K_0^2 (n_1^2 - n_2^2).$$

In the first approximation of dispersion theory in a "nonlinear step" the variation in the complex envelop of the light wave electric field

$$\mathcal{E} = ½\, eR(r)\{E(z, t)\exp[i(\omega t - kz)] + \text{к.с.}\}$$ + complex conjugation}

is described by the equation

$$2i\left(\frac{\partial E}{\partial z} + \frac{1}{v_{\text{гр}}}\frac{\partial E}{\partial t}\right) = kn_{2\,\text{эф}}|E|^2 E,$$

whose general solution takes the form

$$E(z, t) = E_0\left(t - \frac{z}{v_{\text{гр}}}\right)\exp\left[\frac{i}{2} kzn_{2\,\text{эф}}\left|E_0\left(t - \frac{z}{v}\right)\right|^2\right], \tag{10}$$

where $E_0(t - z/v)$ is an arbitrary pulse envelope given in the $z = 0$ cross-section.

As indicated by (10) the nonlinearity of the refractive index of the optical fiber material causes frequency sweeping in the pulse

$$\omega(t) = \omega_0 + \Delta\omega(t),$$
$$\Delta\omega(t) = -\frac{1}{2n_0} kzn_{2\,\text{эф}} \frac{\partial}{\partial t}\left|E_0\left(t - \frac{z}{v}\right)\right|^2. \tag{11}$$

Therefore during the propagation of a powerful ultranarrow pulse in an optical fiber the pulse spectrum $E_0^2 \sim \exp(2t^2/\tau_0^2)$ is broadened:

$$\delta\omega = \frac{\partial \omega}{\partial t}\,\tau_0 \simeq \frac{2kzn_{2\,\text{эф}}E_0^2}{n_0\tau_0}, \tag{12}$$

which becomes equal to the initial spectral pulse width $\delta\omega_0 = 2/\tau_0$ over a nonlinear length $z_{\text{НЛ}}$:

$$z_{\text{нл}} = n_0\,(kn_{2\,\text{эф}}E_0^2)^{-1}.$$

It is well known that as an FM pulse propagates in a linear dispersive medium (i.e., over the next "dispersion step" on z) its temporal envelope is distorted as a result of the conversion of FM into AM. We will introduce a quantitative measure of the dispersion in the "dispersion step" $\Delta z_{\text{Д}}$:

$$k(\omega) = k_0 + \frac{\partial k}{\partial \omega}\bigg|_{\omega_0}(\omega - \omega_0) + \frac{1}{2}\,\frac{\partial^2 k}{\partial \omega^2}\bigg|_{\omega_0}(\omega - \omega_0)^2,$$

where

$$\frac{\partial^2 k}{\partial \omega^2} = -\frac{1}{v_{\text{гр}}}\,\frac{\partial v_{\text{гр}}}{\partial \omega} = \frac{1}{c^2}\frac{\lambda^3}{2\pi}\,\frac{\partial^2 n_{\text{гр}}}{d\lambda^2}.$$

In the second approximation of dispersion theory the amplitude of the wave packet in the nth dispersion layer will be determined by the solution of the linear parabolic equation

$$2i\left(\frac{\partial E}{\partial z} + \frac{1}{v}\frac{\partial E}{\partial t}\right) = -\frac{\partial^2 k}{\partial \omega^2}\,\frac{\partial^2 E}{\partial \tau^2},$$

$$E(z,t) = (2\pi i k''_\omega z)^{-1/2}\int_{-\infty}^{\infty} E_0(\eta)\exp\left[-i\frac{(\tau-\eta)^2}{2k''_\omega z}\right]d\eta.$$

In the particular case of Gaussian pulse propagation with linear frequency modulation

$$E(z=0,t) = E_0\exp\left[-t^2/\tau_0^2 + i\xi t^2\right]$$

the solution takes the form

$$E(z=\Delta z, t') = E_1\exp\left(-t'^2/\tau_1^2 + i\xi' t'^2 + i\zeta\right).$$

Here

$$t' = t - \Delta z/v_{\text{гр}};\quad E_1 = E_0\sqrt{\tau_0/\tau_1};\quad \xi' = \frac{2k''_\omega \Delta z}{\tau_1^2\tau_0^2} + \frac{\xi\tau_0^2}{\tau_1^2};$$

$$\tau_1(\Delta z) = \tau_0\left[(1 + 2\xi k''_\omega\Delta z)^2 + 4(\Delta z)^2/z_{\text{Д}}^2\right]^{1/2}; \tag{13}$$

$z_{Д} = \tau_0^2/k''_\omega$ is the dispersion length over which the pulse is broadened by a factor of two.

As we see from (13) in the absence of frequency modulation in the entrance pulse, dispersion will cause pulse broadening which will depend on its initial duration τ_0. Due to the linearity of the problem the dispersion will not change the spectral width of the pulse. Temporal pulse broadening will be accompanied by the appearance of frequency modulation in the pulse

$$\frac{\partial\omega}{\partial t} = 2\xi' = \frac{4k''_\omega \Delta z}{\tau_0^2 \tau_1^2 (\Delta z)}, \tag{14}$$

where the sign of $\partial\omega/\partial t > 0$.

As indicated by (11) and (13) the frequency modulation accumulated over the "nonlinear" layer $\Delta z_{НЛ}$ distorts the temporal envelope over the "dispersion" step $\Delta z_{Д}$.

The physical picture of the change occurring in an optical pulse with linear frequency modulation $\omega(t) = \omega_0 + \frac{\partial\omega}{\partial t} t$ is most easily observed in the dispersion (linear) layer $\Delta z_{Д}$. Dispersion of the envelope velocity $\partial v_{гр}/\partial\omega$ causes the various spectral components in such a pulse to travel at different envelope velocities. If the signs of the frequency modulation and the dispersion are such that the frequency in the pulse increases in time (i.e., $\partial\omega/\partial t > 0$), while $\partial v_{гр}/\partial\omega < 0$, the spectral components with the higher frequencies will lag those of lower frequency and the pulse will expand in time during the propagation process. On the other hand, if the condition $\frac{\partial\omega}{\partial t}\frac{\partial v_{гр}}{\partial\omega} > 0$, is satisfied, during the propagation process the higher frequencies will overcome the lower frequencies and the pulse will be compressed in time.

Thus, in its sequential passage through the layers of an optical fiber with purely nonlinear and purely dispersive properties for $\Delta z \to 0$ the combined influence of the reactive nonlinearity $n_2|E|2$ and the dispersion of the envelope velocity $\partial v_{гр}/\partial\omega$ may cause self-steepening of the optical pulse. The criterion for such nonlinear propagation of pulses in silican optical fibers is $\partial^2 k/\partial\omega^2 < 0$. Since the dispersion properties of the optical fibers are determined, as demonstrated in section 1, not only by the dispersion of the optical material of the optical fiber but also by the modal (waveguide) dispersion, selection of the specific refractive index profile will make it possible to effectively control the sign and magnitude of the parameter $\partial^2 k/\partial\omega^2$ (see section 1):

$$\frac{\partial^2 k}{\partial\omega^2} = -\frac{\lambda^2}{2\pi c} S_1,$$

where $S_1 = -\frac{\lambda}{c}\frac{d^2 n_э}{d\lambda^2}$.

If the frequency shifts in the pulse arising in the "nonlinear" and "dispersion" layers cancel out $z\ \frac{\partial\omega}{\partial t}\tau_0\Big|_{\Delta z_{нл}}, \frac{\partial\omega}{\partial t}\tau_0\Big|_{\Delta z_{д}}$, so-called soliton propagation conditions are established in which there is no frequency modulation in the pulse and the pulse waveform remains unchanged throughout the optical fiber length.

From the condition

$$\frac{\partial\omega}{\partial t}\tau_0\Big|_{\Delta z_{нл}} = \frac{\partial\omega}{\partial t}\tau_0\Big|_{\Delta z_{д}}$$

and relations (12) and (14) we may determine the range of critical parameters $E_{кр}$ and $\tau_{кр}$ in which it is possible to implement soliton pulse propagation conditions:

$$E^2_{кр}\,\tau^2_{кр} = \frac{2n_0 k''_\omega}{kn_{2\ эф}}. \tag{15}$$

Thus, in the negative ("anomalous") envelope velocity dispersion range $k'_\omega < 0$ ($\lambda > 1.3$ μm) the joint action of dispersion and nonlinearity may result in self-steepening of the pulses and the formation of optical solitons. We note that minimizing the parameter $k''_\omega \sim S_1$ (see section 1) makes it possible to significantly reduce the critical powers $P_{кр}$ of soliton formation (15).

A strict theoretical analysis of optical pulse propagation in optical fibers accounting for the dispersion and nonlinearity of the refractive index may be carried out within the scope of a nonlinear parabolic equation for quasi-monochromatic wave packets [15, 16] by generalization to the weakly-inhomogeneous channeling media of the so-called nonlinear Schrodinger equation

$$i\frac{\partial E}{\partial z} = -\frac{1}{2}\frac{\partial^2 k}{\partial\omega^2}\frac{\partial^2 E}{\partial\tau^2} + \alpha k\frac{n_2}{n_0}|E|^2 E - i\gamma E. \tag{16}$$

Here $\tau = t - z/v_{гр}$ is "traveling" time; $v_{гр} = \partial\omega/\partial k$ is the group velocity; $k = k_0 n_{эф}$ is the propagation constant of the channeled mode; γ is the absorption coefficient.

It is convenient to carry out a transformation to dimensionless variables in equation (16) by normalizing the length to the dispersion length $z = z/z_{д}$, and the amplitude to the characteristic initial value of E_0 at the pulse peak and time to the initial pulse duration τ_0:

$$i\frac{\partial\psi}{\partial z} = \frac{1}{2}\frac{\partial^2\psi}{\partial\tau^2} + R|\psi|^2\psi - i\gamma\psi. \tag{17}$$

The soliton solution of (17) takes the form ($R \equiv 1$)

$$\psi(z, \tau) = \kappa_0 \operatorname{sech}(\kappa_0 \tau) \exp\left(-\frac{i\kappa_0^2 z}{2}\right).$$

The primary characteristics of optical solitons include peak power $P_{\text{кр}}$ corresponding to a fundamental soliton and the pulsation period of the waveform of multisoliton pulses z_0:

$$P_{\text{кр}} = \frac{\lambda_0 A_{\text{эф}}}{4 n_2 z_0}, \quad z_0 = 0{,}322 \frac{\pi^2 c^2 \tau_0^2}{D(\lambda)\lambda_0}. \tag{18}$$

Here τ_0 is the pulse duration at half amplitude; λ_0 is the wavelength in the vacuum; c is the speed of light; $D(\lambda)$ is the parameter determining the group velocity dispersion. Expressions (18) are written in the form proposed by Stolen and Mollenauer [22, 23].

3. The Nonlinear Dynamics of the Temporal Envelope and Spectrum of Optical Solitons in Extended Optical Fibers with Linear Energy Dissipation

The first and central problem in the development of long-range fiber-optic soliton links has been the problem of compensating nonlinear signal distortions in the optical fiber caused by optical emission losses.

Unlike linear fiber-optic communications links in which the data transmission speed is limited by the dispersion of the group velocity of the launched pulses, with soliton pulse propagation in nonlinear fiber-optic communications links the data transmission speed will be limited by the total optical losses in the optical fiber and the peak power of the transmitted signals.

The mathematical model describing soliton dynamics in an optical fiber with linear energy dissipation is based on the nonlinear Schrodinger equation

$$i\frac{\partial \psi}{\partial z} = \frac{1}{2}\frac{\partial^2 \psi}{\partial \tau^2} + |\psi|^2 \psi - i\gamma\psi, \tag{19}$$

where the linear optical loss parameter over a dispersion length of $\gamma = \Gamma\tau_0^2/|\partial^2 k/\partial\omega^2|$ is included in the right half.

We know [16] that (19) when $\gamma \equiv 0$ is a completely integrated model, and its isolated waves are true solitons. Here the total pulse energy (determining the soliton "mass") and the Hamiltonian of equation (19)

$$\mathcal{E} = \int_{-\infty}^{\infty} \psi\psi^* d\tau, \quad H = \int_{-\infty}^{\infty} \left(\frac{\partial\psi}{\partial\tau}\frac{\partial\psi^*}{\partial\tau} - |\psi|^4\right) d\tau$$

are integrals of motion.

Within the framework of the modified nonlinear Schrodinger equation (19) the quantities $\mathcal{E}$ and H satisfy the following equations:

$$\frac{\partial \mathcal{E}}{\partial z} = -2\gamma \mathcal{E}, \tag{20}$$

$$\frac{\partial H}{\partial z} = -2\gamma H + 2\gamma \int_{-\infty}^{\infty} |\psi|^4 d\tau. \tag{21}$$

It is easy to obtain from (20) and (21) a solution describing a "quasi-soliton" with a slowly varying amplitude and duration:

$$\psi(z, \tau) = \kappa(z) \operatorname{sech}(\kappa(z)\tau) e^{i\varphi(z)}, \tag{22}$$

where the functions $k(z)$ and $\varphi(z)$ take the form

$$\kappa(z) = \kappa_0 \exp(-2\gamma z), \quad \varphi(z) = \frac{\kappa_0^2}{8\gamma}(1 - e^{-4\gamma z}) z.$$

A solution of the type (22) was obtained within the framework of perturbation theory in study [13] and was used to estimate the maximum data transmission speed in long optical fibers. As indicated by (22) in this case the rate of data transmission $M \sim 1/\tau(z) \sim \exp(-2\gamma z)$ diminishes exponentially with distance. However since the solution of (22) obtained in perturbation theory ($\gamma z \ll 1$) is valid only on limited pulse propagation paths, the issue of the structural stability criterion of the soliton in the transmission link arises.

We will determine the concept of soliton structural stability in a transmission link (an optical fiber). If in propagating along an optical fiber the functional waveform of the pulse is determined by a hyperbolic secant function independent of the optical losses γ, we will take such a pulse to be structurally-stable. Structurally-stable will also refer to the N-soliton solutions of nonlinear Schrodinger equation (19) that do not change their functional form on paths corresponding to the soliton period $z_0 = (\pi/2)z_{\text{д}}$.

We will determine the quantity $\gamma z_{\text{кр}}$ representing the structural stability criterion of quasi-soliton solutions of the modified nonlinear Schrodinger equation (19) where the relative difference in the pulse dispersions calculated within the framework of perturbation theory and with exact numerical integration of (19) is equal to the initial pulse dispersion at the entrance to the optical fiber:

$$\sigma(\gamma z_{\text{кр}})_{\text{тв}} - \sigma(\gamma z_{\text{кр}})_{\text{числ}} = \sigma(z=0).$$

As indicated by numerical calculations [20] the parameter $\gamma z_{\text{кр}}$ monotonically diminishes with growth of optical losses γ and amounts to $\gamma z_{\text{кр}} = 0.76$, 0.66, and 0.56 for $\gamma = 0.04$, 0.06, and 0.08, respectively.

Thus even with $\gamma \leqslant 0.03$ significant distortions at one soliton period arise in the soliton waveform and spectrum that differ from the "quasi-soliton" solution. These aberrations in the temporal envelope and spectrum accumulate with distance, so that when $\gamma z \gg \gamma z_{кр}$ solution (22) does not correctly describe pulse behavior in the general case. Therefore it is not proper to select a value of 20 km as the repeater length (corresponding to a parameter $\gamma z \sim 1$), since in this case aberrations in soliton form will accumulate in each repeater section.

Recently the issue of the possible application of multisoliton pulses in nonlinear fiber-optic communications links has been widely discussed: so-called breezers or bions [18-22]. The propagation of an N-soliton pulse $\psi(0, \tau) = N\text{sech}\tau$ in a nonlinear dispersive medium with zero losses is described by nonlinear Schrodinger equation (19). The characteristic form of the solution of the nonlinear Schrodinger equation for different γ and $n = 3$ is shown in Fig. 2.

Fig. 3. gives the calculated dependencies of the mean squares of the pulse duration $\langle\tau^2(z)\rangle$ on the optical fiber length. As we see from Figs. 2 and 3 the primary contribution to the mean-square duration of an N-soliton pulse does not come from the narrow central peak, but rather from the broad pedestal whose energy increases with growth of losses γ. In this case over significant pulse propagation paths the degree of energy localization in the path drops significantly compared to a single-soliton pulse. In Fig. 3 the dashed curves represent the dynamics of the mean square of the duration $\langle\tau^2(z)\rangle$ of a single-soliton pulse calculated for identical γ parameters. The existence of separate optical fiber sections in which the dispersion of multisoliton pulses exceeds the dispersion of single-soliton pulses and the irregularity in the restoration of the temporal structure of the pulses along the optical fiber length clearly present additional difficulties in the development of nonlinear fiber-optic communications links based on N-soliton pulses. Moreover, the decay of bound soliton states that may occur over sufficiently long nonlinear pulse propagation (see Fig. 2c) may generally cause total loss of information. This effect could have been present in experiments [14] to reduce the initial pulse duration to 4 ps.

One possible method of reducing the role of the optical loss factor in nonlinear soliton fiber-optic communications links may be converting to the medium IR [23]. In this range a number of materials (crystals and glasses) may have optical losses of 10^{-2}-10^{-3} dB/km, which is one to two orders of magnitude lower than in silicate glasses. The most progress to date has been achieved in the development and fabrication of optical fibers based on fluoride glasses [23]. On the other hand picosecond pulse sources have been extensively developed in this same spectral range ($\eta \sim 2$-3 μm).

The difference in the spectral dependence of the group velocity dispersion causes a significant change in the soliton formation conditions in the fluoride glass-based optical fibers compared to vitreous silican optical fibers. If in the vitreous glasses the primary soliton parameters - critical power $P_{кр}$ and soliton period z_0 - satisfy the relations ($A_{эф} \sim \lambda^2$)

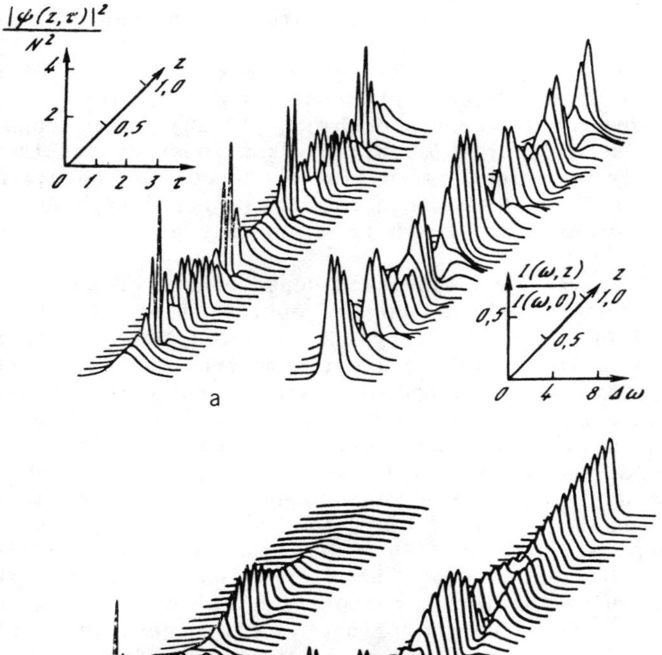

Fig. 2. The dynamics of the temporal envelope and spectrum of the bound state $N = 3$ of solitons in an optical fiber calculated for losses of $\gamma = 0$ (a), $\gamma = 0.5\ \gamma_{Kp}$ (b), $\gamma = \gamma_{Kp}$ (c), $\gamma_{Kp} = 0.26$, $\gamma_{Kp} = (-2/\pi)\ln(1 - 1/N)$

$$P_{\text{кр}} \sim [\lambda^3(\lambda-\lambda_0)]/\tau_0^2, \qquad z_0 \sim \tau_0^2/(\lambda-\lambda_0),$$

in fluoride-glass-based optical fibers the $z_0(\lambda, \tau)$ and $P_{\text{кр}}(\lambda, \tau)$ relations change significantly:

$$P_0 \sim \lambda^6/\tau^2, \quad z_0 \sim \tau^2/\lambda^3.$$

Therefore the conversion to the medium infrared causes a significant increase in the peak power necessary for soliton formation. Thus, for example, with $A_{\text{эф}}$ = 100 μm^2, λ = 3 μm the condition for soliton propagation of a pulse in a fluoride-glass-based optical fiber takes the form

$$P\tau^2 > 2 \text{ kwt}\cdot\text{ps}^2.$$

Moreover, the optical losses in such optical fibers will not exceed the critical value:

$$2\tau^2\gamma_{\text{кр}} = \frac{\ln N^2}{z_0} = 2{,}6\cdot10^{-4}\ln N^2\ (\text{cm}^{-1}\cdot\text{ps}^2),$$

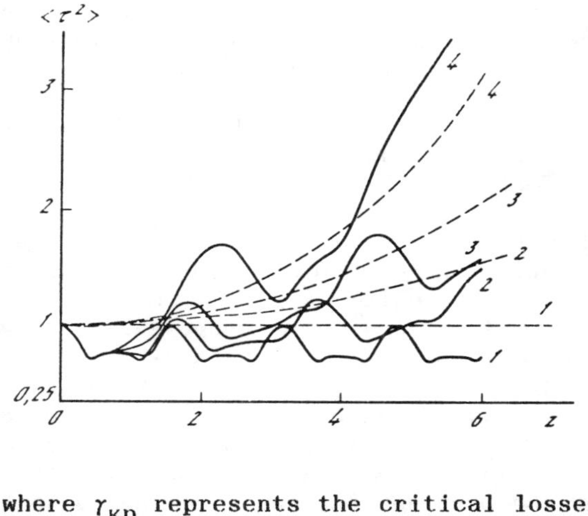

Fig. 3. The mean square pulse duration Nsechτ plotted as a function of optical fiber length for γ = 0.0, 0.02, 0.04, and 0.08 (curves 1, 2, 3, 4) and N = 3
The dashed curves represent the change in the mean square duration of a single-soliton pulse calculated with identical losses

where $\gamma_{\text{кр}}$ represents the critical losses in the optical fiber where a single-soliton pulse is formed from an N-soliton pulse at a single period z_0.

In concluding this section we will consider the influence of nonlinear losses in an optical fiber on the soliton propagation dynamics. Here we will limit our examination to the case where the perturbation to the soliton waveform is sufficiently small.

Pulse propagation in an optical fiber accounting for linear and nonlinear absorption effects is described within the framework of the nonlinear Schrodinger equation

$$i \frac{\partial \psi}{\partial z} = \frac{1}{2} \frac{\partial^2 \psi}{\partial \tau^2} + |\psi|^2 \psi - i\gamma_\Pi \psi - i\gamma_2 |\psi|^2 \psi - i\gamma_c |\psi_c|^2 \psi, \tag{23}$$

in which γ_Π, γ_2, and γ^C are the linear, two-photon and stimulated Raman scattering-absorption coefficients, respectively. The solution of (23) may be written as a quasi soliton with a slowly varying amplitude, duration and phase

$$\psi(z, \tau) = \kappa(z)\text{sech}[\kappa(z)\tau] \exp \frac{i}{2} \int_{-\infty}^{z} \kappa^2(z')\, dz', \tag{24}$$

$$\kappa^2(z) = \frac{\kappa^2(0)\exp(-Az)}{1 - \frac{\kappa^2(0)}{A}[B - \exp(-Az)\,(B + Cz)]}, \tag{25}$$

$$A = 4(\gamma_\Pi + \gamma_c I_c(0); \; B = \frac{8}{3}\gamma_2 + C/A; \; C = \frac{8}{3}\gamma_c^2 I_c^2(0),$$

where $I_C(0)$ is the intensity of spontaneous Raman scattering in the optical fiber. As indicated by (24) and (25) when $\gamma_2 = \gamma_C = 0$ solution (25) describes a quasi soliton with exponential broadening from propagation in the optical fiber. Nonlinear absorption in soliton optical fibers may be considered one of the primary factors limiting data transmission speed in the conversion to the medium IR.

4. The Nonlinear Dynamics of Soliton Gain in Optical Fibers. The "Skipping" Pump Technique

The specifics of optical soliton gain are related to the fact that the soliton duration is unambiguously related to the peak power (see section 2). Different methods of soliton amplification in nonlinear fiber-optic communications links have been proposed. Study [24] has considered the possibility of optical soliton gain by injecting into the optical fiber a plane wave of the same frequency and phase as the soliton. In this method of signal amplification a portion of the plane wave energy is transmitted to the soliton, although here it is necessary to develop "refined" methods of phase and frequency detection and to destroy the portion of plane wave energy that remains unused during the amplification process. Another method of amplifying solitons is proposed in studies [25] which developed linear soliton gain techniques in superconductor laser repeaters. One possible method of solving this problem is to make the conversion to the medium IR [23]. Rivlin examined stimulated Raman signal gain in study [26] for a linear fiber-optic communications link with quasi- compensated losses that combines the function of transmitting medium and distributed gain amplifier.

The idea for the possible application of stimulated Raman scattering to attenuation compensation and soliton generation in optical fibers was first proposed in study [27] and was later developed in studies [28-31]. However the subject in studies [27, 30] was limited to an approximation of a given pump field and stationary Raman gain conditions, i.e., the exhaustion of the pump wave was ignored together with the mismatch between the group velocities of the pump pulses and the Stokes emission.

Unlike studies [27, 30] the present study proposes and develops a "skipping pump" method of soliton amplification in nonlinear fiber-optic communications links employing nonstationary stimulated Raman scattering in single-mode optical fibers. Lugov has developed a mathematical model of the stimulated Raman process in a single-mode optical fiber that accounts for the emission intensity dependence of the refractive index [32] in a first approximation of dispersion theory. In the second approximation accounting for the dispersion of the group velocities within the spectrum of each of the interacting waves the nonlinear dynamics of stimulated Raman soliton propagation in an optical fiber is described by a system of Schrodinger equations for the complex pump wave envelopes $E_H(z, \tau)$ and the first Stokes component $E_C(z, \tau)$:

$$2i\left[\frac{\partial E_H}{\partial z} + \left(\frac{1}{v_H} - \frac{1}{v_c}\right)\frac{\partial E_H}{\partial \tau}\right] = -\frac{\partial^2 k}{\partial \omega^2}\frac{\partial^2 E_H}{\partial \tau^2} + \frac{n_2 k_H \alpha}{n_0}|E_H|^2 E_H +$$

$$+\frac{n_2 k_H \alpha}{n_0}|E_c|^2 E_H - ig\frac{\omega_H}{\omega_c}\alpha Q E_c - i\gamma_H E_H,$$

$$2i\frac{\partial E_c}{\partial z} = -\frac{\partial^2 k_c}{\partial \omega^2}\frac{\partial^2 E_c}{\partial \tau^2} + \frac{n_2 k_c \alpha}{n_0}|E_c|^2 E_c + \frac{n_2 k_c \alpha}{n_0}|E_H|^2 E_c +$$

$$+ ig\alpha Q^* E_H - i\gamma_c E_c, \quad \left(T_2\frac{\partial}{\partial \tau} + 1\right)Q = E_H E_c^*.$$

Here $E_{H,C}(z, \tau)$ are the complex amplitudes of the interacting waves; Q is the amplitude of the molecular vibration wave; $\tau = t - z/v_C$ is "traveling" time related to the Stokes pulse; $v_{H,C}$ are the group velocities of the pulses; g is the gain of the Stokes pulse at the peak of the stimulated Raman scattering line; $\gamma_{H,C}$ are the linear absorption coefficients; T_2^{-1} is the half-width of the spontaneous Raman scattering line.

We will assume that the duration of a soliton propagating in an optical fiber at the Raman frequency satisfies the condition $\tau_C > T_2$ (for vitreous silica glasses $T_2 \sim 10^{-13}$ s). Moreover, since the modulation instability of the pump wave in the range of negative dispersion of group velocity $k''_{\omega^2<0}$ may cause a significant distortion in its temporal structure and a decay into separate pulses [15], we will assume that the entire spectrum of the pump wave is concentrated in the range $\partial^2 k_H/\partial\omega^2 > 0$, at the same time that the spectrum of the amplified pulse is located in the spectral range $d^2 k_C/d\omega^2 < 0$.

We will consider the gain of a Raman soliton in a plane unmodulated pump wave field, assuming the relation $|\psi_C|^2 \ll |\psi_{H0}|^2$ to be satisfied over the gain interval $z \ll 1$, where we may also ignore the mismatch of the group velocities of the pulses at the fundamental and Raman frequencies. In this case the evolution of the Raman soliton is described by the following nonlinear Schrodinger equation:

$$i\frac{\partial\psi_c}{\partial z} = -\frac{1}{2}\frac{\partial^2\psi_c}{\partial\tau^2} + |\psi_c|^2\psi_c + ig\exp(-2\gamma_H z)\psi_c - i\gamma_c\psi_c, \tag{26}$$

where $g = G|\psi_{HO}|^2$ is the stationary gain increment; γ_H are linear losses at the pump frequency over the dispersion length.

It is easy to obtain a solution of equation (26) describing a quasi soliton with a slowly varying amplitude and phase:

$$\psi(z, \tau) = \eta(z)\mathrm{sech}[\eta(z)\tau]\exp(i\varphi(z, \tau)).$$

Here the function $\eta(z)$ takes the form

$$\eta(z) = \eta(0)\exp\left[2g\left(\frac{1-\exp(-2\gamma_H z)}{2\gamma_H}\right) - 2\gamma_H z\right]. \tag{27}$$

Relation (27) describes the competition of the linear damping effect and the stimulated Raman gain effect during soliton propagation through the optical fiber.

We will examine the soliton restoration process. Assume that by the launch time of the pump wave into the optical fiber the soliton duration is determined by the expression

$$\eta(z_c) = \eta(0)\exp(-2\gamma_c z_c).$$

Then when $\gamma_H z \ll 1$ over the optical fiber length

$$z_p = z_c/(g/\gamma_c - 1) \tag{28}$$

the waveform of the initial pulse is recovered. As we see from (28) when $g \gg \gamma_c$ the restoration length of the Raman soliton is $z_p \ll z_c$.

Fig. 4 (curves 1) show the change in soliton energy $\mathcal{E}$, its amplitude $|\psi(z, 0)|$ and duration τ in an optical fiber with linear losses $\gamma = 0.02$ with pump wave injection $|\psi_{HO}|^2 = 100$ at time t_1 determined from the condition $\mathcal{E} = 1$ over the interval in z $\Delta z = 0.01$ ("impact" gain). Fig. 4 (curves 2) show the recovery of the soliton parameters for the case of distributed gain $g\exp(-2\gamma z) = 2\gamma$ beginning at time $t = t_1$. Unlike study [30] in the numerical experiments the dynamics of Raman soliton gain were also considered in the pulsed pump field.

A pulse at the Stokes frequency in a pulsed pump field will be amplified more efficiently if the group velocities of the interacting pulses coincide. Here the energy center of the Stokes pulse does not shift during the gain process. It is possible to compensate the detuning of the group velocities of the pump and Raman pulses either

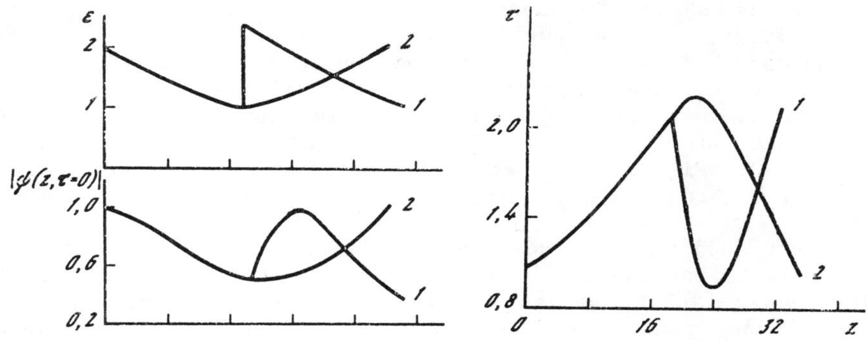

Fig. 4. The dynamics of Raman soliton gain by "impact" pumping (curves 1) and distributed gain (curves 2)

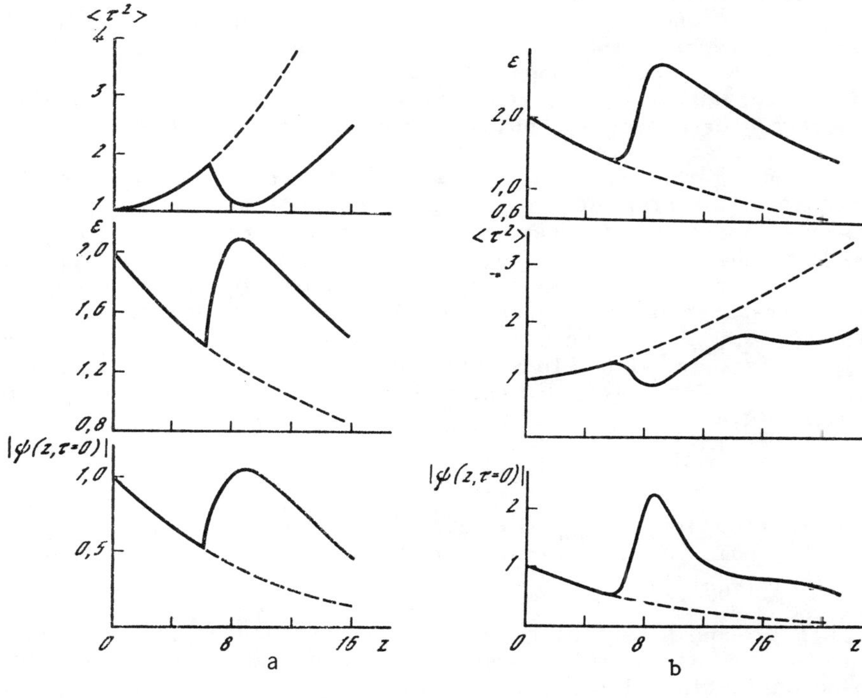

Fig. 5. Raman soliton gain in a pulsed pump field in the spectral ranges $\partial^2 k_H/\partial\omega^2 > 0$ (a) $M^2 k_H/\partial\omega^2 < 0$ (b)

by appropriate selection of the refractive index profile or by using pump and Raman pulses at wavelengths symmetrical with respect to the "zero" dispersion of the group velocity.

Fig. 5a shows the change in the mean square duration $<\tau^2>$ of Raman solitons, their energy $\mathcal{E}$ and intensity $|\psi(z, 0)|^2$ with distance for the case of launching of a pump pulse of the type $\psi_{H0}(0, \tau) = \text{sech}\tau$ at time t_1 determined from the condition $|\psi(z, \tau = 0)|^2 = 0.5$ ($z = 6$). The dashed curve represents the quantities $<\tau^2>$, $\mathcal{E}$, $|\psi(z, 0)|^2$ with no pump pulse. The pump pulse is located in the spectral range $\partial^2 k_H/\partial\omega^2 > 0$. Fig. 5b gives analogous relations calculated for a pump pulse of $\psi_H = \text{sech}\tau$ in the spectral range $M^2 k_H/\partial\omega^2 < 0$.

Study [31] unlike study [30] has proposed and developed a soliton gain technique for nonlinear fiber-optic communications links that employs nonstationary Raman gain in single-mode optical fibers: the "skipping" pump technique. The essence of this method is that in conjunction with the launch of the high-power information pulse - the optical soliton of the signal envelope - a low-intensity pump wave is launched into the optical fiber; this wave falls within the spectral range corresponding to positive ("normal") group velocity dispersion ($\partial v_H/\partial\lambda > 0$). When $v_H > v_c$ the pump wave travels more rapidly than the soliton at the Stokes frequency and consequently the latter is continuously within the range of powerfully excited molecular vibrations of the medium. As a result the Raman soliton contains all the pump energy stored over the pulse period of the pulse train.

By tuning the pump emission wavelength it is possible to change the relative detuning of the group velocities of the pulses at the fundamental and Raman frequencies and thereby vary the rate of energy exchange between the interacting pulses over the coherence interaction length $z_{\text{кор}} = \tau_{H0}/(\tau_{c0} c)$, where τ_{H0} is the duration of the pump pulse. If total transfer of pump energy to Raman soliton energy occurs over length $z_{\text{вз}} = z_{\text{кор}}$, the condition for recovery of the Raman soliton in an optical fiber with linear optical losses over the interaction length $z_{\text{вос}}$ is determined by the following relation between the primary parameters:

$$\tau_{H0}/(\tau_{c0}c) = (-1/2\gamma)\ln[2/(2 + |\psi_{H0}|^2\tau_{H0}/\tau_{c0})].$$

The interaction length $z_{\text{вос}}$ is limited by the stimulated Raman scattering threshold in the optical fiber resulting from spontaneous noise $z_{\text{вос}} \ll 16/(g|\psi_{H0}|^2)$.

The proposed "skipping" pump technique makes it possible not only to restore the waveform and energy of pulses in nonlinear fiber-optic communications links but also to achieve nonlinear conversion of solitons and formation of high-energy Raman solitons. As indicated by the calculations the proposed technique may be used in the optimum range of γ parameters to transmit a single-soliton pulse of duration $\tau_{c0} = 3$ ps over a distance $z_{\text{вз}} = 5.5\ z_{\text{д}}$, which equals 15 km for the following parameters: $\lambda_H = 1.235\ \mu m$, $\lambda_c = 1.305\ \mu m$ ($\lambda_0 = 1.27\ \mu m$), $\gamma = 0.2$ dB/km and $|\partial^2 k_c/\partial^2\omega^2| = 3.3\cdot 10^{-29}\ s^2/cm$.

Consequently the Raman effect in optical fibers which until recently was considered a factor that limits the range and speed of data transmission on optical fibers is quite the reverse: the Raman effect may be directly employed to develop high-capacity nonlinear soliton optical communications systems. The direct experimental demonstration of optical Raman soliton recovery in a 10 km optical fiber for the case of pumping from the opposite optical fiber end (opposing stimulated Raman scattering) recently carried out in study [33] has revealed new possibilities for solving the problem of optical soliton retransmission.

5. The Interaction of Optical Solitons

The physical picture of the interaction of solitons of nonlinear Schrodinger equation (19) with zero energy dissipation ($\gamma \equiv 0$) has been investigated most extensively in studies [34-38]. Analogous results were recently obtained in study [35] for application to data transmission by optical solitons. As demonstrated in study [36] the dynamics of weakly-interacting solitons are identical to the dynamics of classical particles with pair interaction potential. With purely real initial conditions of the type

$$\psi(0, \tau) = \operatorname{sech}(\tau - \Delta/2) + \operatorname{sech}(\tau + \Delta/2), \tag{29}$$

which are of primary interest for data transmission on optical fibers it is possible that bound soliton states may exist where the distance between the solitons and their phase shifts vary periodically in time: the solitons will merge periodically from the maximum intersoliton distance of Δ to complete overlapping of the wave fields (Fig 6a), N = 1. The dependence of the oscillation period of the bound soliton pair (29) on the maximum distance between the solitons takes the form

$$p = \frac{\pi}{2} e^{\Delta/2}.$$

Thus nonlinear soliton interaction will cause a complete loss of information at the overlap point between the wave fields: the soliton "merging" length.

Studies [35, 38] have examined various methods of reducing interaction between solitons: introducing a phase shift of π in the initial pulses [35, 38] and launching of solitons of different amplitudes and phases [35].

In the present study we will focus on one additional possibility for controlling the interactive forces between solitons which is manifest from the phase modulation of the initial pulses. We know [39] that regular phase modulation makes it difficult to achieve stationary soliton propagation in optical fibers: the pulse amplitude and duration oscillate with distance out of phase. The fact that the dispersion of a phase modulated pulse oscillates in a limited interval during its nonlinear propagation in an optical fiber makes it possible to propose such a pulse for data transmission on optical fibers.

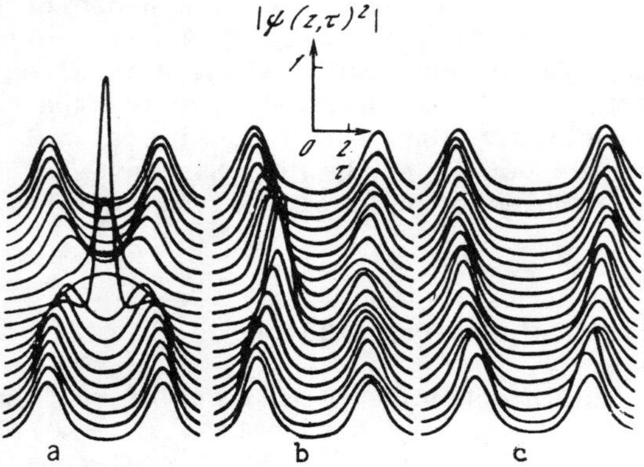

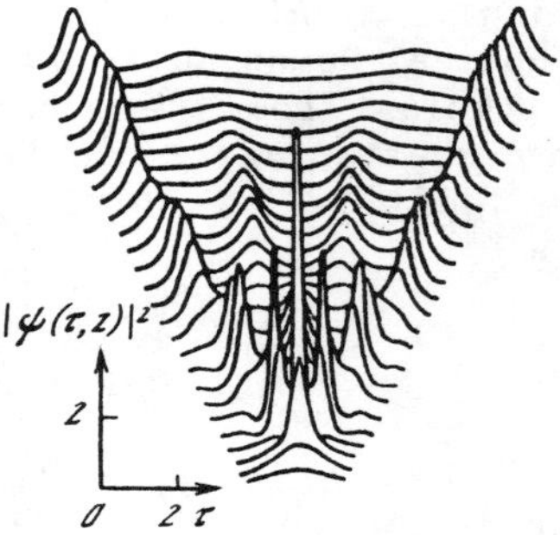

Fig. 6. The dynamics of soliton interaction in a single period p with zero energy dissipation

Fig. 7. The decay of the wave packet under the influence of initial frequency modulation: $\psi(0, \tau) = 5 \operatorname{sech}\tau \exp(i\beta_{\text{кр}}\operatorname{sech}^2\tau)$; $\Delta z = 0.02$; $\beta = -7.32$; $N = 5$

We should emphasize, however, that the frequency modulation level in the pulse is limited by the existence of soliton conditions $H < 0$ (see section 3) and cannot exceed the critical value $\xi < \xi_{\text{кр}}$ [20]. For pulses of the type $\psi(0, \tau) = N \operatorname{sech} \tau \exp(i\xi\operatorname{sech}^2\tau)$ the parameter $\xi^2_{\text{кр}} = 1.09\ (2N^2 - 1)$ [20]. When $\xi > \xi_{\text{кр}}$ the initial packet decays (Fig. 7).

We will consider the dynamics of nonlinear interaction of phase modulated pulses of the following type:

$$\psi(z = 0, \tau) = \operatorname{sech}(\tau - \Delta/2)\exp[i\,\xi\operatorname{sech}^2(\tau - \Delta/2)] + \operatorname{sech}(\tau + \Delta/2)\exp[i\,\xi\operatorname{sech}^2 \times (\tau + \Delta/2)]. \quad (30)$$

With interaction of pulses of the type in (30) the phase mismatch effect and the antiphase oscillations in pulse duration serve to reduce the "attractive forces" between solitons. As demonstrated by numerical calculations there exists for each fixed value of the parameter Δ an optimum pulse frequency modulation rate ξ_{opt} where the soliton trajectories will not undergo significant changes with distance. The characteristic features of soliton interaction calculated for $\Delta = 6$, $N = 1$, $\xi = 0$ (Fig. 6a) and $\xi = \xi_{\text{opt}} = 0.3$ (Fig. 6b) are shown in Fig. 6.

The "attractive forces" of the solitons to the common center of mass are determined by the degree of overlap between the "tails" of the separate solitons. Therefore phase modulation of emission in the "tail" overlap range $\tau = 0$

$$\psi(0, \tau) = [\operatorname{sech}(\tau - \Delta/2) + \operatorname{sech}(\tau + \Delta/2)]\ \exp(i\,\xi\operatorname{sech}^2\tau) \quad (31)$$

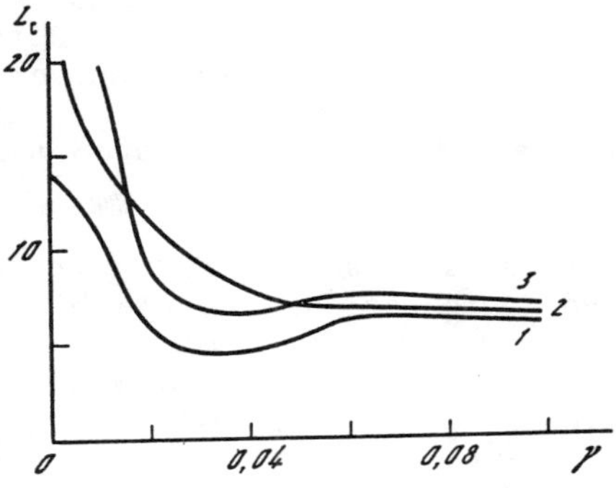

Fig. 8. The "merging" length L_c plotted as a function of linear optical losses in an optical fiber γ calculated for ξ = 0 (1), 0.5 (2), and 1 (3)

Fig. 9. The interaction dynamics of bound soliton states (N = 2) (a) and the maximum intensity of a "noise" pulse plotted as a function of z for ξ = 0 (1), 0.5 (2), and 1 (3) (b)

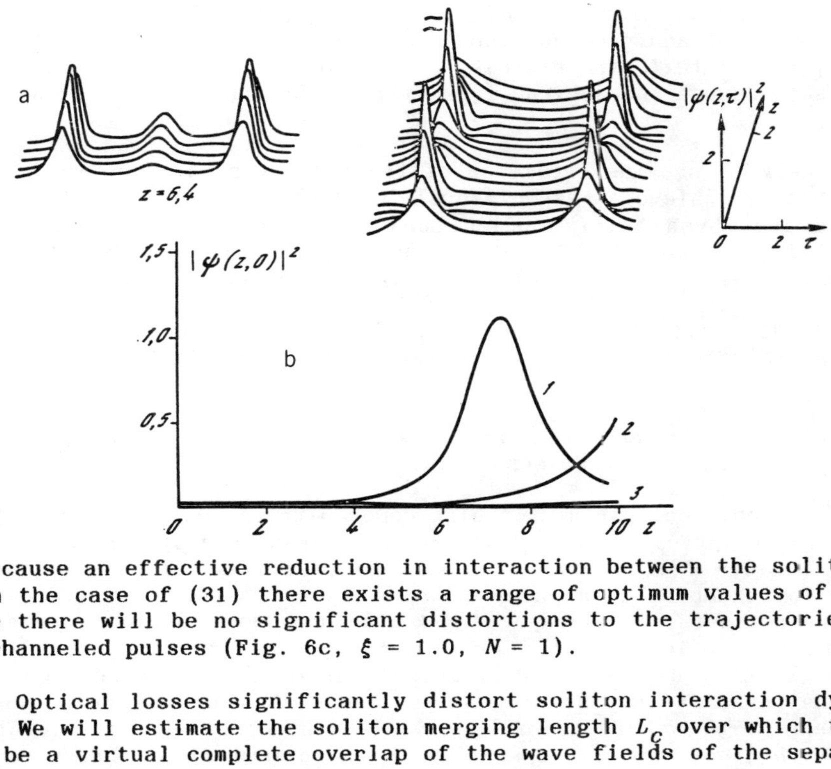

will cause an effective reduction in interaction between the solitons. As in the case of (31) there exists a range of optimum values of ξ_{opt} where there will be no significant distortions to the trajectories of the channeled pulses (Fig. 6c, ξ = 1.0, N = 1).

Optical losses significantly distort soliton interaction dynamics. We will estimate the soliton merging length L_c over which there will be a virtual complete overlap of the wave fields of the separate solitons by means of the following relation:

$$(|\psi_{max}(L_c,\tau)|^2 - |\psi(L_c,\tau=0)|^2)\,/\,|\psi_{max}(L_c,\tau)|^2 \leqslant 0{,}1,$$

where $|\psi_{max}(Lc,\ \tau)|^2$ is the maximum pulse intensity.

Fig. 8 gives the "merging" length L_c plotted against the dimensionless optical losses γ calculated for the initial conditions in (31) and ξ = 0, 0.5, and 1. As indicated by the calculations the "merging" length increases significantly in the parameter range $\gamma \leqslant 0.01$ and $\xi \simeq 1$. With an increase in losses the joint action of the nonlinear pulse blurring effects and pulse interaction will serve to reduce the "merging" length. Therefore phase modulation of the initial pulses is not advantageous for transmitting pulses over significant distances in optical fibers with a loss level of $\gamma > \gamma_{\text{Kp}}$.

In conclusion we will discuss certain aspects of the interaction of multisoliton pulses. As indicated by numerical calculations a characteristic feature of the interaction of N-soliton pulses ($N \geqslant 2$) is the generation of "noise" signals resulting in partial loss of information (Fig. 9). Frequency modulation of the type in (31) makes it possible to suppress the appearance of "noise" signals. Fig. 9 gives calculated dependencies of the maximum noise signal intensity on the nonlinear interaction length of pulses compared to the case of the interaction of nonfrequency modulated pulses. An increase in the frequency scan rate in the vicinity of overlap of the pulse "tails" $\tau = 0$ serves to reduce the maximum intensity of the "noise" signal (see Fig. 9).

We will provide estimates of the primary parameters where it is possible to achieve an experimental implementation of the effects discussed above: an initial pulse duration $\tau_0 = 3$ ps, an optical loss level of $\gamma = 0.01$ ($\Gamma = 0.2$ dB/km) and a group velocity dispersion $D = 15$ ps/(nm·km) a critical soliton formation power $P_{\text{Kp}} = 0.8$ w a dispersion length $z_{\text{д}} = 0.5$ km and a "soliton merging" length without FM of $L_c = 10$ km. With a Pulse period of $T = 6\tau_0$ it is possible to achieve a data transmission speed of $M = 500$ (Gbit/s)·km.

6. The Nonlinear Dynamics of Femtosecond Light Pulses in Optical Fibers

The physical picture of high-power ultranarrow light pulse propagation in optical fibers becomes significantly more complex in the transition to femtosecond durations of the channeled pulses [40-44]. Thus, for example, at a pulse duration $\tau_0 = 0.1$ ps the group velocity dispersion in fused silica at $\lambda = 1.5$ μm broadens a pulse by more than 1 m in the optical fiber. By minimizing the first and second order group velocity dispersion coefficients as described in section 1, in the best case it is possible to increase the pulse dispersion spreading length by two orders of magnitude to hundreds of meters. Therefore with linear pulse propagation in an optical fiber there are fundamental difficulties in achieving data transmission speeds in the order of terabits per second over a kilometer or more. Therefore nonlinear (soliton) propagation of optical pulses in optical fibers may represent the only possibility for the further development of a fiber-optic communications link and for incorporating the range of data transmission speeds measured in terabits per second per kilometer.

We will consider in greater detail the primary physical effects that occur from the propagation of high-power femtosecond light pulses in optical fibers. Accounting for the terms of second order of smallness in the parameter $\theta = T/\tau_0$ (T is the period of the light wave) the propagation of a light pulse is described by the modified nonlinear Schrodinger equation [40]

$$i\frac{\partial\psi}{\partial z}=\frac{1}{2}\frac{\partial^2\psi}{\partial\tau^2}+R|\psi|^2\psi-i\eta\frac{\partial}{\partial\tau}(|\psi|^2\psi),$$

$$R=N^2=a\frac{z_{\text{д}}}{z_{\text{нл}}},\quad z_{\text{нл}}=\frac{n_0}{2kn_2I_0},\quad \eta=R\frac{2n_0}{c\tau_0 k}. \tag{32}$$

The first three integrals of motion of the modified nonlinear Schrodinger equation (32) take the form

$$\widetilde{W}=W=\int_{-\infty}^{\infty}\rho d\tau;\ \widetilde{P}=P-\eta\int_{-\infty}^{\infty}\rho^2 d\tau;$$

$$\widetilde{H}=H-\frac{3}{2}\eta\int_{-\infty}^{\infty}\rho j d\tau+\eta^2\int_{-\infty}^{\infty}\rho^3 d\tau;\ \rho=|\psi|^2,\ j=2\mathrm{Im}\,\psi^*\psi_\tau.$$

The existence of the integrals of motion of equation (32) indicates the possibility for the existence of soliton solutions. Such solutions were found in plasma physics where (32) describes the propagation of Alfvén waves [44]. The primary difference between single-soliton solutions of (32) and the solutions of the nonlinear Schrodinger equation is in the significant frequency modulation of the pulse [40, 45].

As indicated by numerical investigations [45] the existence of a perturbation in the nonlinear Schrodinger equation ($\eta \neq 0$) produces instability in the bound soliton states. They decay into separate pulses. As a result a short intense pulse is generated at the leading edge; this pulse has a duration determined by the duration of the narrow peak against the broad pedestal background that arises within the framework of the unperturbed nonlinear Schrodinger equation model (see section 3). A differentiating feature of the pulse is its RF modulation.

We will provide a qualitative explanation of the decay picture. As noted above the existence of a perturbation in (32) causes a pulse to acquire asymmetric frequency modulation with preferential favoring of the high-frequency portion of the spectrum as it propagates through an optical fiber. The change in sign of frequency modulation from self-steepening of the multisoliton pulse and its subsequent breakdown into separate fragments causes an intense pulse to be generated at the leading edge of the pulse envelope. The subsequent periodic change in sign of frequency modulation and the continuous shifting in the center of gravity of the spectrum of the pulse forming at the leading edge towards higher frequencies will result in its further steepening and "accelerated" motion of the envelope peak at the leading edge. Therefore a stationary RF modulated pulse appears at the leading edge with the maximum of the pulse trailing the motion of the envelope peak. It is easy to obtain expressions for the "velocity" of the center of gravity of the packet $\delta\tau = \frac{1}{W}\frac{d^2}{dz}\int_{-\infty}^{\infty} \tau\rho d\tau$ and its "acceleration" with distance

$$\delta\dot{\tau} = \frac{1}{W}\frac{d^2}{dz^2}\int_{-\infty}^{\infty} \tau\rho d\tau:$$

$$W\,\delta\tau = \frac{1}{2}\int_{-\infty}^{\infty} (j - 3\,\eta\rho^2)d\tau,\ W\,\delta\dot{\tau} = \eta\int_{-\infty}^{\infty} \rho j_\tau\, d\tau.$$

When the conditions $\delta\tau\,|_{z=0} = \delta\dot{\tau}\,|_{z=0} = 0,\ z \ll \frac{4}{3}\frac{1}{\eta^2 R}$ are satisfied it is possible that within the framework of (33) a stationary solution of the following type may exist

$$\psi(z,\tau) = \operatorname{sech}(\sqrt{R}\,\tau)\exp\left[-i\frac{Rz}{2} + i\frac{3}{2}\frac{\eta}{\sqrt{R}}\operatorname{th}(\sqrt{R}\,\tau) - i\frac{3}{8}\eta^2 z\operatorname{sech}^4(\sqrt{R}\tau)\right]. \tag{33}$$

A pulse of the type in (33) may evidently be used to transmit information in an optical fiber by means of femtosecond pulses.

Comprehensive experimental investigations of optical solitons in optical fibers carried out in study [22] have demonstrated that as the power of the channeled pulses is increased, the agreement between experiment and numerical calculations performed within the framework of the nonlinear Schrodinger equation model for $\eta = 0$ is degraded. This discrepancy between theory and experiment may be explained in the modified nonlinear Schrodinger equation model that accounts for the nonstationary of nonlinear polarization. We carried out numerical calculations corresponding to experiments from study [22] for pulses of τ_0 = 6-7 ps, N = 3. In this case the parameter η = 0.00375-0.0066. Fig. 10 gives the pulse waveforms at one soliton period. A comparison of the temporal pulse envelopes calculated for various values of the parameter η reveals that the nonstationarity of polarization infers the recovery of the initial pulse waveform at the soliton period.

The possibility for generating femtosecond pulses in an optical fiber from the decay of the initial multisoliton packets requires accounting for the higher order dispersion effects. In the following approximation of dispersion theory equation (32) will take the form

$$i\frac{\partial\psi}{\partial z} = \frac{1}{2}\frac{\partial^2\psi}{\partial\tau^2} + R\,|\psi|^2\,\psi - i\,\eta\frac{\partial}{\partial\tau}(|\,\psi\,|^2\psi) + i\beta\frac{\partial^3\psi}{\partial\tau^3},\ .$$

where $\beta = \pm\frac{1}{6}\frac{|k'''|}{|k''|}\frac{1}{\tau_0}$ and is related to the dispersion coefficients S_1 and S_2 by the following relations:

$$k''_\omega = \frac{\lambda^2}{2\pi c}S_1,\ k'''_\omega = \frac{\lambda^3}{2\pi^2 c^2}S_1 + \frac{\lambda^4}{4\pi^2 c^2}S_2.$$

The order of the parameter β determines the magnitude of waveform distortion and the change in the group velocity of the pulse [13].

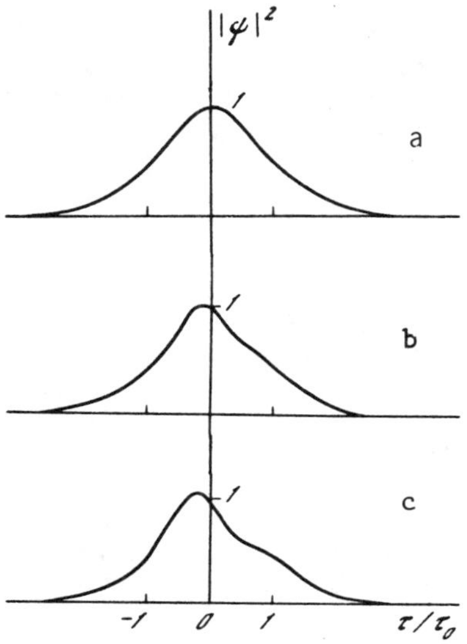

Fig. 10. Distortion to the wave packet envelope ψ = sechτ for $R = 9$ at one soliton period for $\eta = 0$ (a), 0.00375 (b), and 0.0066 (c)

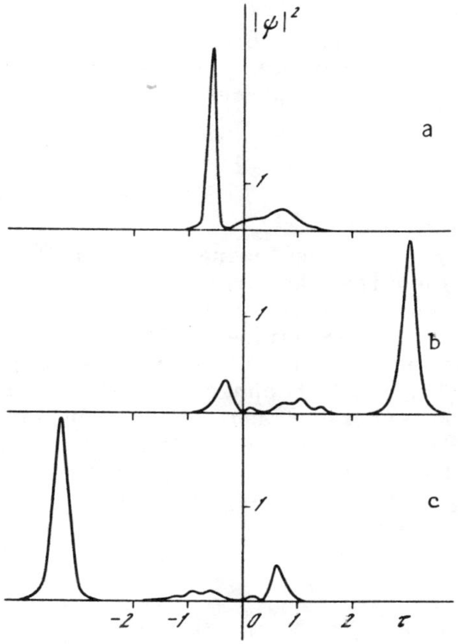

Fig. 11. The "decay" of the wave packet ψ = sechτ for $R = 9$; $z = z_0$; $\eta = 0.1$; $\partial^3 k/\partial\omega^5 \neq 0$
The pulse waveforms are shown sequentially for the following parameters: $\beta = 10^{-3}$ (a), $5\cdot 10^{-2}$ (b), and $5\cdot 10^{-2}$ (c)

In linear (with respect to the field) propagation of an ultranarrow pulse, dispersion of the order k''' ($\eta = 0$) will produce asymmetric distortion to the pulse waveform and will generate oscillating "tails" associated with the Airy function [13]. Depending on the sign of the parameter β the peak of the envelope either is stretched to the trailing edge ($\beta > 0$) or the leading edge ($\beta < 0$) of the pulse, while the oscillating part also ends up either on the trailing edge or the leading edge [5].

In nonlinear pulse propagation the influence of third order dispersion, as indicated by numerical calculations, becomes significant for $\beta > 0.1\,\eta$. Fig. 11 shows the temporal structures of the generated pulses in one period z_0 with "decay" of the packet $\psi(z = 0, \tau)$ = sechτ, $R = N^2 = 9$ calculated for various parameters β.

The dispersion k_ω''' may radically change the decay of the multisoliton pulse. Depending on the sign of the parameter β the intense pulse will be formed either on the leading edge ($\beta < 0$) or on the trailing edge ($\beta > 0$). Moreover significant dispersion spreading of the packet will occur.

Thus, the possibilities for fabricating optical fibers with different ratios of the parameters S_1 and S_2 (k_ω'' and k_ω''') make it possible to implement different nonlinear channeling states of femtosecond pulses.

Conclusion

In reviewing the results derived in the present study we may identify the promise of soliton optical wave propagation in optical fibers for data transmission applications. As demonstrated in this study the unique possibilities for data transmission by optical solitons are particularly strongly manifest compared to traditional "linear" techniques under rapid development at the present time. Recent experiments by Stolen and his associates on Raman gain of solitons in a single-mode optical fiber 10 km in length [33] indicates the possibility for the fabrication of soliton fiber-optic communications links in the near future.

Bibliography

1. Linke, R.A., Kasper, B.L., Campell, J.C., et al. "120 km lightwave transmission experiment at 1 Gbit/s using a new long-wavelength avalanche photodetector" ELECTRON. LETT., 1984, V. 20, No. 12, p. 498-499.

2. Dianov, Ye.M., Kuznetsov, A.A. "Spectral channel multiplexing in fiber-optic communications links: (a survey)" KVANTOVAYA ELEKTRON, 1983, V. 10, No. 2, p. 245-263.

3. Kapron, F.P., Keck, D.E. "Pulse transmission through a dielectric optical waveguide" APPL. OPT., 1971, V. 10, No. 10, p. 1519-1523".

4. Unger, Kh.G. "Planarnye i volokonnye opticheskie volnovody" [Plane and fiber-optic waveguides] Translation from the English, edited by: V.V. Shevchenko, Moscow: Mir, 1980, 656 p.

5. Marcuse, D. "Pulse distortion in single-mode fibers" APPL. OPT., 1980, V. 19, No. 10, p. 1653-1660.

6. Gloge, D. "Dispersion in weakly guiding fibers" Ibid, 1971, V. 10, No. 11, p. 2442-2445.

7. Belanov, A.S., Dianov, Ye.M. "Maximum data transmission speeds on optical fibers" RADIOTEKHNIKA, 1982, V. 37, No. 2, p. 35-43.

8. Bhagavatula, V.A. "Segmented-core single-mode fibers with low loss

and low dispersion" ELECTRON. LETT., 1983, V. 19, No. 9, p. 317-318.

9. Najenski, A., Unger, H.G. "Single-mode W-fibers with profile threshold for low dispersion over a wide spectral range" AEU, 1983, V. 37, No. 9/10, p. 336-338.

10. Belanov, A.S , Dianov, Ye.M., Krivenkov, V.I. "Optical fibers with stable emission polarization" KVANTOVAYA ELEKTRON., 1984, V. 11, No. 6, p. 1273-1275.

11. Belanov, A.S., "Investigation of multilayered optical fibers": Dissertation for Doctorate of Physics and Mathematics, 01.04.04 Moscow, 1980, 282 p.

12. Hasegawa, A., Tappert, F. "Transmission of stationary nonlinear optical pulses in dispersive dielectric fibers. I: Anomalous dispersion" APPL. PHYS. LETT., 1973, V. 23, No. 3, p. 142-144.

13. Hasegawa, A., Kodama, J. "Signal transmission by optical solitons in monomode fiber" IEEE PROC., 1981, V. 69, p. 1145-1150.

14. Mollenauer, L.F., Stolen, R.H., Gordon, J.P. "Experimental observation of picosecond pulse narrowing and solitons in optical fibers" PHYS. REV. LETT., 1980, V. 45, No. 13, p. 1095-1098.

15. Litvak, A.G., Talanov, V.I. "The application of a parabolic equation to calculation of fields in dispersive nonlinear media" IZV. VUZOV. RADIOFIZIKA" 1967, V. 10, p. 539-548.

16. Kakharov, V.Ye., Shabat, A.B. "Exact theory of two-dimensional self-focusing and one-dimensional self-modulation in nonlinear media" ZhETF, 1971, V. 61, No. 1(7), p. 118-134.

17. Tomlinson, W.J., Stolen, R.H., Shank, C.V. "Compression of optical pulses chirped by self-modulation in fibers" J. OPT. SOC. AMER. B. 1984, V. 1, No. 2, p. 139-149.

18. Blow, K.J., Doran, N.J. "Solitons in optical communications" IEEE J. QUANT. ELECTRON., 1983, V. 19, p. 1883-1888.

19. Sisakyan, I.N., Shvartsburg, A.B. "The nonlinear dynamics of picosecond pulses in optical fibers" KVANTOVAYA ELEKTRON., 1984, V. 11, No. 9, p. 1703-1721.

20. Dianov, Ye.M., Nikonova, Z.S., Serkin, V.N. "The dynamics of soliton bound state formation and decay in optical fibers" Preprint of the Institute of General Physics of the Academy of Sciences of the USSR, No. 13, Moscow, 1985, 11 p.

21. Makhan'kov, V.G. "Solitons and numerical experimentation" FIZIKA ELEMENTAR. CHASTITS I ATOM. YADRA, 1983, V. 14, No. 1, p. 123-180.

22. Stolen, R.H., Mollenauer, L.F., Tomlinson, W.J. "Observation of pulse restoration at the soliton period in optical fibers" OPT. LETT., 1983, V. 8, No. 3, p. 186-188.

23. Dianov, Ye.M., Prokhorov, A.M., Serkin, V.N. "The prospects for soliton generation in optical fibers in the medium IR" DOKL. AN SSSR, 1983, V. 273, Nol. 5, p. 112-1115.

24. Hasegawa, A., Kodama, J. "Amplification and reshaping of optical solitons in glass fiber. I" OPT. LETT., 1983, V. 7, No. 6, p. 285-289.

25. Hasegawa, A., Kodama, J. "Amplification and reshaping of optical solitons in glass fiber. II, III" Ibid, V. 8, p. 342-344.

26. Rivlin, L.A. "An optical transmission link with quasi-compensated losses" KVANTOVAYA ELEKTRON., V. 3, No. 7, p. 1611-1612.

27. Isaev, S.K., Kornienko, L.S., Kravtsov, N.V., Serkin, V.N. "Ob odnoy vozmozhnosti generatsii solitonov" [One possibility for soliton generation] XI Bsesoyuz. konf. po kogerent. i nelineynoy optike (22-25 noyab. 1982 g.): Tez. dokl. [The 11th All-Union Conference on Coherent and Nonlinear Optics (22-25 Nov. 1982): Topic paper] Erevan, 1982, V. 2, p. 470-471.

28. Vysloukh, V.A., Serkin, V.N. "The generation of high-energy stimulated emission solitons in optical fibers" PIS'MA I ZhETF, 1983, V. 38, No. 4, p. 170-172.

29. Vysloukh, V.A., Serkin, V.N. "Nonlinear transformation of solitons in optical fibers" IZV. AN SSSR. SER. FIZ., 1984, V. 48, No. 9, p. 1777-1781.

30. Hasegawa, A. "Numerical study of optical soliton transmission amplified periodically by the stimulated Raman process" APPL. OPT., 1984, V. 23, No. 19, p. 3302-3309.

31. Dianov, Ye.M., Nikonova, Z.S., Prokhorov, A.M., Serkin, V.N. "The nonlinear dynamics of soliton Raman gain in fiber-optic communications links" DOKL. AN SSSR, 1985, V. 283, No. 6, p. 1342-1346.

32. Lugovoy, V.N. "Stimulated Raman emission and frequency scanning in an optical waveguide" ZhETF, 1976, V. 71, No. 4, p. 1307-1319.

33. Mollenauer, L.F., Stolen, R.H., Islam, M.N. "Experimental demonstration of soliton propagation in long fibers loss compensated by Raman gain" OPT. LETT., 1985, V. 10, No. 5, p. 229-231.

34. Karpman, V.I., Solovev, V.V. "A perturbational approach to the two-soliton systems" PHYSICA 3D, 1981, V. 87, p. 596-603.

35. Gordon, J.P. "Interaction forces among solitons in optical fibers" OPT. LETT., 1983, V. 8, No. 11, p. 596-598.

36. Gorshkov, K.A., Ostrovskiy, L.A. "Soliton interaction in unintegrated systems (the direct perturbation method and its applications" Preprint of the Institute of Applied Physics of the Academy of Sciences of the USSR, Gor'kiy, 1981, 19 p.

37. Blow, K.J., Doran, N.J. "Bandwidth of nonlinear (soliton) optical communications systems" ELECTRON. LETT., 1983, V. 19, p. 429-430.

38. Hermansson, B., Yevick, D. "Numerical investigation of soliton interaction" Ibid. p. 570-571.

39. Fattakhov, A.M., Chirkin, A.S. "The nonlinear propagation of phase-modulated light pulses" KVANTOVAYA ELEKTRON., 1984, V. 11, p. 2349-2355.

40. Anderson, D., Lisak, S. "Nonlinear asymmetric self-phase modulation and self-steepening of pulses in long optical waveguides" PHYS. REV. A, 1983, V. 27, p. 1393.

41. Gustafson, T.K., Taran, J.P., Haus, A., et al. "Self-modulation, self-steepening, and spectral development of light in small-scale trapped filaments" PHYS. REV., 1969, V. 177, No. 1, p. 306-313.

42. Yang, G., Shen, Y.R. "Spectral broadening of ultrashort pulses in a nonlinear medium" OPT. LETT., 1984, V. 9, No. 11, p. 510.

43. Wadati, M., Konuo, K., Ichikawa, Y.H. "A generalization of inverse scattering method" IPPJ-381, Nagoya Univ., Japan, 1979.

44. Kawata, T., Sakai, J., Kobayashi, N. "Inverse method for the mixed nonlinear Schrodinger equation and soliton solutions" J. PHYS. SOC. JAP., 1980, V. 48, No. 4, p. 1371-1379.

Mechanical Strength of Long-Length Vitreous Silica Glass Optical Fibers

V.A. Bogatyrev, M.M. Bubnov, N.N. Vechkanov
A.N. Gur'yanov, S.L. Semenov

Abstract: A fabrication technology for high-strength optical fibers is developed from systematic measurements of the breaking strength of optical fibers fabricated by means of chemical deposition from the gas phase in conjunction with an analysis of the causes behind the failure of such optical fibers. Using rod preforms of domestic manufacture a strength of 5.5 GPa was achieved for meter-long optical fiber segments, and a strength of 1.4 GPa was achieved for kilometer optical fibers. The parameter n_T characterizing the rate of flaw growth in optical fibers is measured. It is demonstrated that a 25 year service life of the fibers is guaranteed with a 99% probability if the operational stress acting on the otpcal fibers is three times smaller than the preliminary winding stress.

Introduction

Initially the mechanical strength of optical fibers was not considered among the primary characteristics of optical fibers such as optical losses or bandwidth. Therefore investigations of the mechanical strength of optical fibers began to appear only four or five years after the fabrication of the first low-loss optical fibers when it became clear that the mechanical strength of such optical fibers in the final analysis would determine their suitability for further practical application in optical communications systems and would determine the service lives of such systems. A differentiating feature of the mechanical strength as a performance characteristic of an optical fiber is the fact that the critical factor for the entire construction length is not the average mechanical strength but rather is the minimum strength. Moreover low-loss or narrow bandwidth optical fibers still may find application in internal-facility communications links, while low-strength optical fibers have no practical applications.

According to various estimates of the interatomic bond of the silicon and oxygen atoms the mechanical strength of vitreous silica glass is 10-5 GPa [1, 2], although the actual mechanical strength of vitreous silica fibers is significantly lower. This is due to the fact that the silica glass, particularly its surface, contains a large quantity of microcracks, foreign inclusions and other flaws. When a tensile force is applied to the silica glass the stresses are concentrated at the apexes of the microcracks and therefore the glass sample will begin to break down long before the applied load reaches the theoretical strength of the glass. The concentration of stresses at the crack apex is characterized by the stress intensity factor K [3]:

$$K = Y \sigma \sqrt{a}, \tag{1}$$

where σ is applied stress; a is microcrack depth; Y is the coefficient dependent on the crack geometry (for a semicircular crack $Y = 1.24$). The stress level σ_{Kp} at which failure occurs may be determined from the expression

$$\sigma_{\text{Kp}} = K_{\text{Kp}}/Y\sqrt{a}, \tag{2}$$

where K_{Kp} is the material constant (for vitreous silica glass K_{Kp} = 0.789 $MN/m^{3/2}$ [3]).

Expression (2) makes it possible to estimate the maximum possible flaw size in glass for a given mechanical strength value. For example, in order to provide a mechanical strength of 5 N in a standard optical fiber, there can be no flaw greater than ~2 μm on either the glass surface or the glass bulk. If we compare the size of a single flaw with the total surface area of a 1 km optical fiber (0.4 m^2) it becomes clear that the problem of fabricating optical fibers with even a moderate mechanical strength is a rather difficult process and in order to solve this problem it is necessary to improve the optical fiber fabrication technology to eliminate the causes of flaws greater than that specified in (2).

Fabricating High-Strength Optical Fibers

We developed a horizontal tensile testing machine for optical fiber mechanical strength tests [4]. The roller-type connectors used to fasten the ends of the optical fiber samples in the machine made it possible to eliminate clamping failures of the optical fibers and to investigate their strength up to stresses of the order 6 GPa.

In order to determine the causes behind optical fiber failure both optical fiber ends at the failure site were subjected to microscopic analysis. A "mirror zone" arises around the flaw responsible for optical fiber failure [5]. This zone borders on the "fog zone" showing minor surface roughness which becomes the "fan zone": a region with significant surface inhomogeneities (Fig. 1). The position and dimensions of the mirror zone made it possible for us to determine not only the location but also the size of the flaw responsible for failure. For example, the "mirror zone" diameter for vitreous silica glass is approximately ten times the flaw size [6].

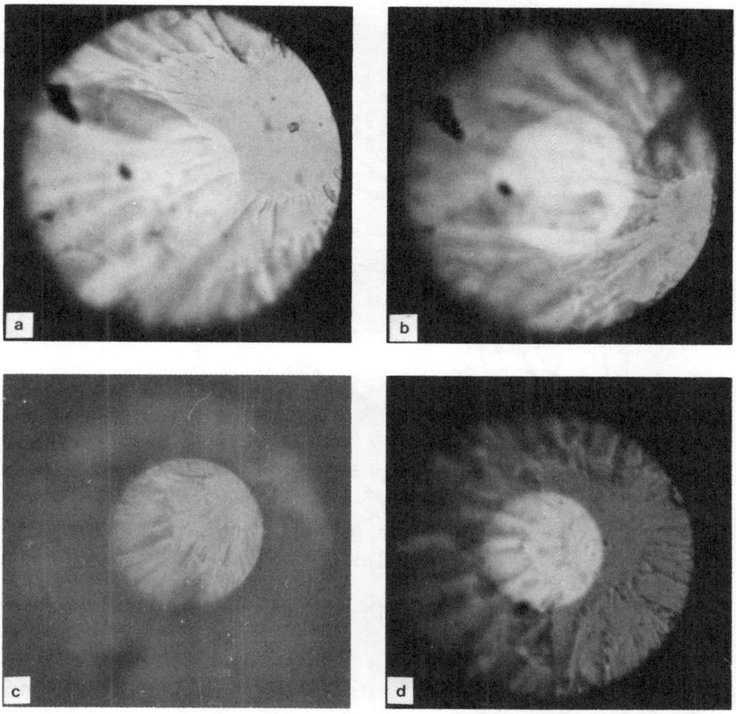

Fig. 1. End surface of optical fibers at the failure sites
a - inclusion in the preform rod; b - surface flaw; c - surface flaw due to uneven coating; d - inclusion on the inner surface of the preform rod

In the initial stages of our research the most common cause of failures were surface flaws in the optical fibers. We developed a wide range of technological operations to eliminate such flaws. An oxygen-propane burner was used to polish the preforms immediately prior to the drawing process. The burner conditions were selected so that the 100 μm surface layer of the preform would evaporate after passed through the burner three times. Immediately after this process the preform was fastened in the clamps of the tensile testing machine and the optical fiber was drawn. Our oxygen-hydrogen burner developed for optical fiber drawing provided sterile preform heating conditions and low (approximately 30 g) tensile force on the fiber during the drawing process. A polymer coating was directly applied to the fiber surface during the drawing process in order to eliminate surface faults. Fig. 2 shows the measurement results produced by the tensile strength machine for our fabricated optical fibers with different polymer coatings.

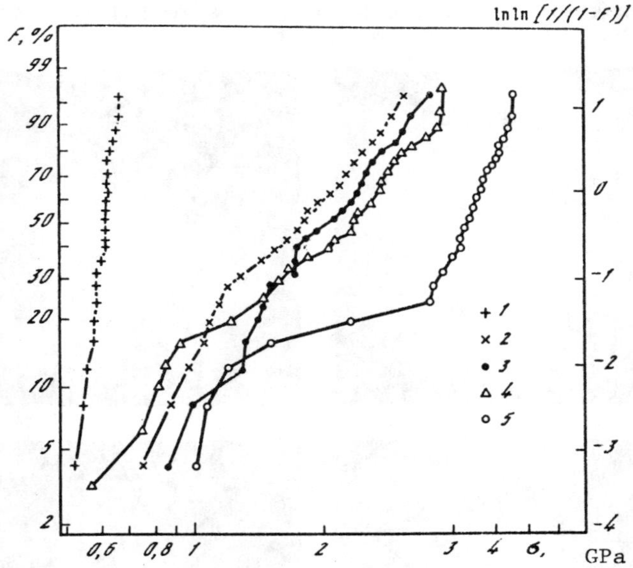

Fig. 2. Weibull graphs of the unit strength distribution function of KU-1 synthetic vitreous silica glass fibers with different polymer coatings
1 - FPKh-3A; 2 - Alston; 3 - AD-9113; 4 - PAI-1; 5 - SIEL 159-167

Since the flaws are randomly distributed along the optical fiber length and are of random size, the strength of the optical fibers may be analyzed statistically. Normally the Weibull function $F(\sigma)$ is used to characterize the strength of optical fibers; this function represents the probability of optical fiber failure at stresses below σ [7]. In Fig. 2 the stresses are plotted on the X-axis while the Weibull functions or the percentage of samples that fail at the given stress level are plotted on the Y axis. In this series of experiments in order to more clearly reveal the dependence of optical fiber strength on the polymer coating material type KU-1 synthetic ultrapure vitreous silica glass bars were used as the preforms; this brand of vitreous silica glass has very few internal flaws. A variety of polymer coatings, primarily lacquer coatings, were directly applied to the fibers drawn from these preforms during the actual drawing process (fiber diameter: 150 μm). As we see from Fig. 2 the strongest fiber sample was coated with a silicone elastomer coating. The strength was improved further by increasing the thickness of this elastomer from 20 to 60 μm (Fig. 3).

Therefore using flame polishing of the initial bars, a sterile heating source for the drawing process and by applying a protective polymer coating of appropriate thickness during the drawing process it was possible to increase the strength of synthetic ultrapure vitreous silica glass fibers to 5-6 GPa. However the strength of fibers fabricated by chemical deposition from the gas phase using the same technological techniques remains low. An analysis of the ends of such optical fibers at their failure sites has revealed that the cause of failure are flaws within the preform rod material. Optical fibers fabricated by chemical deposition from the gas phase consist of approximately 70% preform rod material, while the vitreous silica glass preforms currently manufactured by domestic industry are fabricated from a mineral basic material and therefore the quantity of impurities and foreign inclusions may very over a wide range.

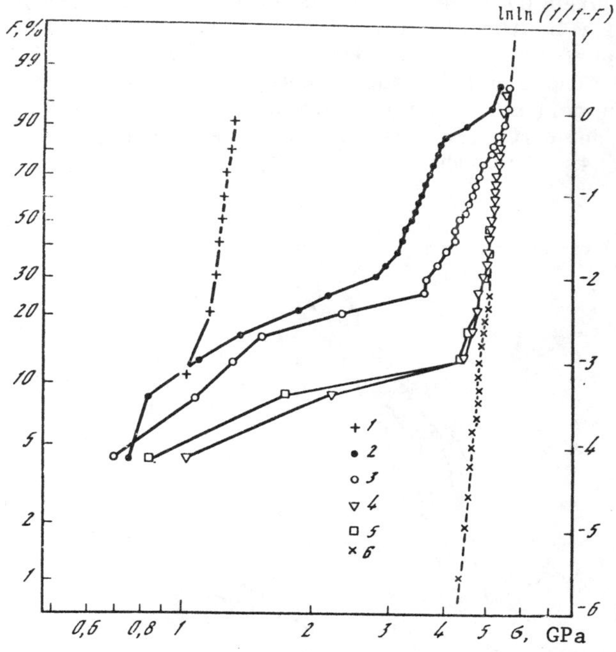

Fig. 3. Weibull graphs of the unit strength distribution function of KU-1 synthetic vitreous silica glass fibers with SIEL 159-167 silicone coating of various thicknesses (μm)
1 - 0; 2 - 20; 3 - 30; 4 - 50; 5 - 60; 6 - dual-layer coating

In order to establish the influence of preform rod quality on the optical fiber strength rod samples fabricated by different technologies at ten domestic and foreign organizations were employed. Results from strength measurements for fibers drawn from these rods are given in Fig. 4. As clearly indicated by the diagram, our test rods could be divided into two groups which for simplicity we shall henceforth call "flawed" and "flaw-free" preform rods.

An X-ray microanalyzer analysis of the element composition of inclusions on the "Camebax" assembly has revealed that in the "flawed" rods more than 90% of the inclusions causing failures contain molybdenum while the remaining inclusions take the form of calcium and iron. It was not possible in certain cases to determine the composition of the inclusions. We may only conclude that the ordinal number of the elements comprising the inclusion is below 11. It is possible that these inclusions are graphite particles. The "flaw-free" rods were similar to the KU-1 synthetic vitreous silica glass in terms of inclusion size and quantity. In these rods the inclusions contained calcium, iron and titanium. As in the "flawed" rods it was not always possible to determine the composition of the flaws.

These investigations have revealed that the strength of optical fibers fabricated by chemical deposition from the gas phase is determined primarily by the quality of the silica preform rods. When using "flaw-free" preform rods the strength of the optical fiber approaches the ultimate strength of homogeneous vitreous silica glass fibers, as the change in concentration of germanium dioxide in the optical fiber core, as we demonstrated previously in study [8] has no significant influence on its strength.

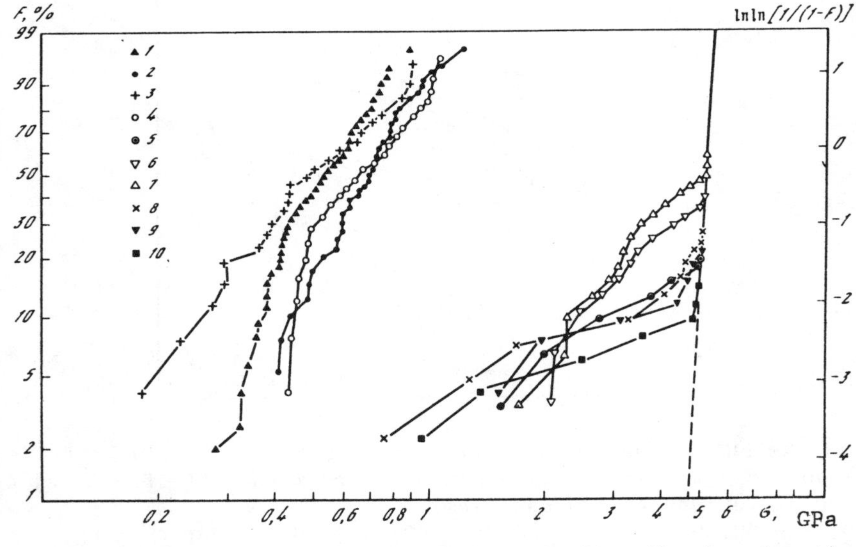

Fig. 4. Weibull graphs of the unit strength distribution function of vitreous silica glass fibers fabricated from various preform rods 1 - Japan; 2-4 - USSR; 5 - German Democratic Republic; 6 - People's Republic of Bulgaria; 7 - USSR; 8 - Federal Republic of Germany; 9- USSR; 10 - KU-1 (USSR)

Mechanical Strength of Long-Length Optical Fibers

By testing optical fiber segments 0.5-2 m in length on the tensile testing machine it was possible to obtain information on the statistics of flaws most often encountered in optical fibers, while the integrity of an optical fiber in the manufacture and operation of an optical cable is determined by the largest flaw along the optical fiber length. The natural method of establishing its true size and therefore the optical fiber strength is to run a strength test of the entire line structure. Clearly in order to determine the statistics of the appearance of such rare flaws it would be necessary to test tens of kilometers of optical fibers or 10^4-10^5-meter segments which is, generally speaking, a difficult process. Therefore, in order to determine the statistics of the appearance of large yet rare flaws that determine the strength of long-length optical fibers we ran the optical fibers through a winding process at different tensile levels. Approximately 150 km of optical fibers 125 μm in diameter were run through this process at tensile force levels of 3, 5, 8, 10, and 15 N. During the winding from one standard bobbin to another the optical fiber crossed two rollers used to create axial tension without slippage. This tension was applied to the optical fiber for 1-2 s. At the same time the drawing force on the optical fiber from the feed bobbin and the winding force of the optical fiber into the receiving

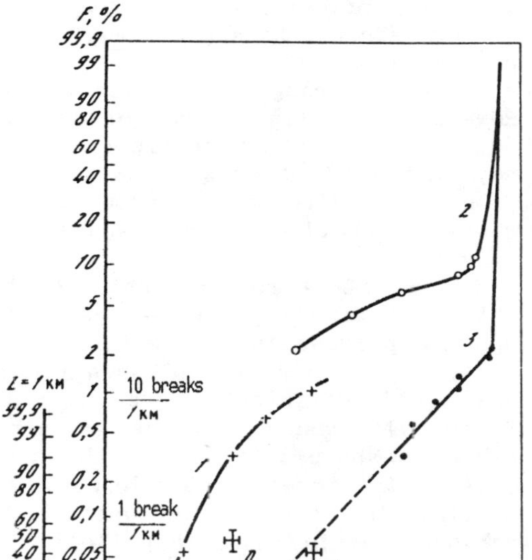

Fig. 5. Weibull graphs of the unit strength distribution function of optical fibers based on results from winding under axial tension

bobbin did not exceed 0.5 N. As indicated by our estimates and after repeated winding processes in these loading conditions there was no significant reduction in optical fiber strength.

The results from winding 150 km optical fibers are shown in Fig. 5 (curve 1) in Weibull coordinates. In drafting this curve the number of optical fiber breaks at a given load was converted to the probability of failure of an optical fiber segment 1 m in length, while the load during the winding process was converted to the equivalent tensile force from dynamic measurements at a relative elongation of the sample of 10%/min. The figure also shows measurement results for the breaking strength of optical fibers for 1 m samples (curve 2). Accounting for the fact that this distribution was obtained for separate typical optical fibers while curve 1 reflects the probability of various flaws appearing along the entire test length of the optical fibers (150 km) the graphs are in good agreement and together reflect the probability of failure of our fabricated optical fibers under various loading conditions.

A fractographic analysis of the optical fiber ends at the breakage sites in the winding process revealed three primary types of flaws responsible for these failures: 1) surface flaws caused by contact between the optical fiber and a foreign particle; 2) microcracks along the optical fiber surface; 3) internal foreign inclusions.

The statistics of optical fiber failures in the winding process have revealed that depending on the winding force approximately 30-55% of all breaks may be attributed to surface cracks and 30-40% of the remaining breaks may be attributed to optical fiber contact with a foreign particle. The surface cracks result from insufficient flame polishing of the preform or a breakdown of the preform or optical fiber surface during the drawing process resulting from contact between the optical fiber and the wall of the internal burner channel or the surface of the support rollers. In addition certain optical fiber breaks could be attributed to the breakdown of the primary coating apparently during the application of the secondary polyamide coating. However the primary cause of surface microcracks today may be attributed to asymmetric application of the primary coating. Such flaws may easily be detected from a fractographic analysis of the optical fiber break site: the "window" at the optical fiber end is on the edge where the coating crack is smallest (see Fig. 1c).

The rheologic properties of the SIEL 159-167 silicone elastomer which is normally used as the primary coating make it impossible to apply a polymer layer greater than 55-65 μm in thickness in a single pass. As we have established previously (see Fig. 3), such a thickness is insufficient. Even a small (10-20 μm) shift of the optical fiber with respect to the geometric center of the polymer coating causes a sharp increase in the number of breaks in the winding process. We therefore developed a technology for applying dual-layer coating directly to the optical fiber in the drawing process. A polycarbonate, fluoroplast [Teflon], or polyacrylic lacquer layer approximately 10 μm in thickness was applied as the second coating above the silicone elastomer. This made it possible for us to significantly reduce the quantity and size of the surface flaws (Fig. 5, curve 3). Hence as the drawing technology improves the strength of long-length optical fibers will be determined by the quantity and size of flaws and foreign inclusions in the preform rods.

At present we employ two types of preform rods fabricated from natural silica by gas melting and electrothermal melting in a vacuum for fabrication of the optical fibers. If we account for optical fiber breaks attributable only to internal flaws, the break probability is shown in Fig. 5 by region *AB* for rods of gas-fused silica glass and by region *CD* for rods of electrothermal-fused silica glass. Optical fibers 1 km in length fabricated from the first group of rods will withstand a winding force of 15 *N* with a probability of 50%. The dashed line passing through region *AB*, point *E* and passing into curve 3 represents the ultimate strength we have achieved so far for optical fibers. As we see from Fig. 5 the yield of intact optical fibers 1 km in length is 85% at a winding force of 0.7 GPa and 50% at a winding force of 1.2 GPa. Optical fibers fabricated from preform rods of the second type had 5-10 times greater the number of internal flaws and therefore lower mechanical strength.

In analyzing the breakage sites the internal inclusions responsible for the failure of the optical fibers during the winding process were normally located in the preform rod material. However in examining industrially manufactured optical fibers we also discovered flaws on the internal surface of the preform rod (see Fig. 1d). The presence of such flaws indicates an insufficient degree of preliminary purification of the preform rods before the final manufacturing process. At the same time we did not observe any flaws responsible for optical fiber failure in the winding process that were located in the optical fiber core.

Our investigation of the strength of long-length optical fibers has revealed that the primary cause of optical fiber failure in the winding process are surface flaws at present. Eliminating the causes of such flaws will make it possible to significantly increase the yield of intact 1 km optical fibers at the same winding tension or with an increased winding tension. However at high tensile winding levels the yield of optical fibers greater than 1 km in length will be limited by the quantity and size of the inclusions in the preform rods.

The Influence of Water on the Mechanical Strength of Optical Fibers

From the viewpoint of the practical application of fiber-optic communications links it is very important to know the nature of variations in optical fiber strength over time. We know that when a load significantly lower than the ultimate load is applied to glass samples for a sufficiently long period they fail. This phenomenon called the static fatigue of glass is due to the growth of internal flaws, while the rates of growth of internal and surface flaws are determined by different mechanisms. When the optical fiber is exposed to tensile stresses the internal flaws expand due to fluctuations in the thermal motion of the glass atoms [1]. The growth of surface flaws is significantly accelerated by water present in the atmosphere; the plastic coatings on standard optical fibers do not offer sufficient protection against moisture. By the beginning of our investigations the data on the influence of water on optical fiber strength published in the literature were contradictory [9-12]. The discrepancy between the experimental results from different authors may be attributed to the fact that the influence of water on optical fiber mechanical strength is largely determined by the drawing technology of the optical fibers as well as the compositions and application techniques of the polymer coatings, i.e., it is necessary to determine parameters characterizing the rate of failure of such optical fibers for the various optical fiber manufacturing production lines.

Using the horizontal tensile testing machine [4] we carried out an investigation of the influence of water on the mechanical strength of our fabricated optical fibers.

The studies were performed on fibers 125 μm in diameter and 1 m in length fabricated from KU-1 synthetic vitreous silica glass. A dual-layer polymer coating (first a silicone elastomer coating 60 μm in thickness and then a 10 μm lacquer layer) was applied directly to the vitreous silica glass fibers during the drawing process by means of an oxygen-hydrogen burner. Fig. 6 presents plots in Weibull coordinates of the unit strength distribution functions of the initial fiber, fibers submerged in water with zero stress for 90 days, and fibers dried in a furnace at 100° after 90 day exposure to water. A comparison of results from these three measurements reveals that after a three-month storage in water the mechanical strength of the vitreous silica glass fibers was reduced by approximately 15%. However after furnace drying the mechanical strength of the fibers was restored to 98% of their initial strength.

Such a reversible nature of the change in the breaking strength indicates that the influence of water on an unloaded fiber does not result in flaw growth. The reduction in the breaking strength of a vitreous silica fiber in water is caused by the change in the dynamics of crack growth during the loading of the fiber on the tensile testing machine.

The following relation is normally used to describe the crack growth rate in glass [11]

$$da/dt = AK^n, \tag{3}$$

where A and n are constants characterizing the dynamics of crack growth; a and K are the same as in expression (1).

Using expressions (1) and (3) we may easily demonstrate that when different operating loads σ_1 and σ_2 are applied to the same optical fiber, the corresponding service lives of the optical fiber τ_1 and τ_2 will satisfy the following ratio:

$$\tau_1/\tau_2 = (\sigma_2/\sigma_1)^n. \tag{4}$$

If we know the breaking strength under short-term loading and using expression (4) we may determine the service life of optical fibers under low operating loads. The parameter n has a significant influence on the results of these estimates. For example, it follows from (4) that if the optical fiber withstands a load of σ_0' for 1 s, under a load of $\sigma_0/2$ it will fail after 9 hours if n = 15 and after 34 years if n = 30.

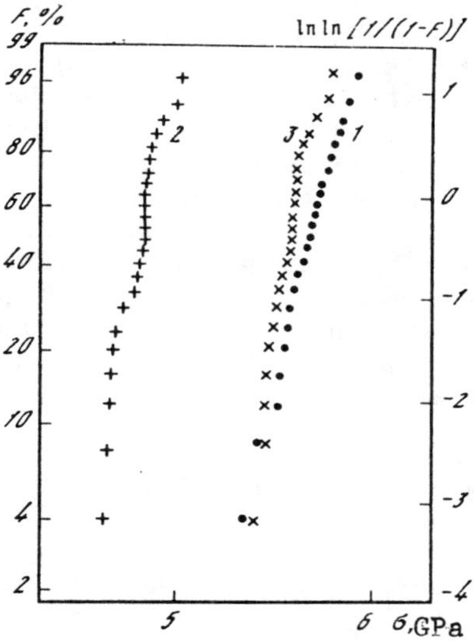

Fig. 6. Weibull graphs of the unit strength distribution function of vitreous silica fibers plotted against the environmental conditions
1 - in air in laboratory conditions; 2 - after exposure to water for 3 months; 3 - after exposure to water for 3 months and subsequent drying at 100°C for 8 hours

Fig. 7. Weibull graphs of the unit strength distribution function of vitreous silica glass fibers at various stress rates
1 - in air; 2 - in water; the rates of relative elongation of the optical fibers are shown in the graph

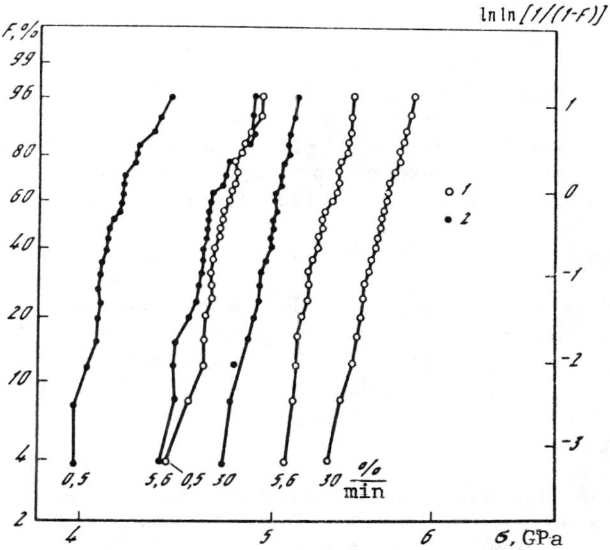

The literature provides values of the parameter n from 15 to 50 [12] and therefore we measured this parameter for our fabricated optical fibers. These measurements were carried out by means of dynamic testing which involved investigating the dependence of the breaking load on the optical fiber stress rate.

As we see from Fig. 7 graphs of the failure probabilities of optical fibers exposed to water are shifted towards the lower stress range, although the distances between these graphs with different stress rates remain the same as for the initial optical fibers. Therefore statistical processing of the measurement results yielded an identical value of the parameter $n = 21.5 \pm 1.5$ for both series of measurements. The value of the parameter A for optical fibers exposed to water for five days was 17 ±3 times smaller than the value of this parameter for the initial optical fibers. Further exposure of the optical fibers to water does not appreciably change the constant A. The service life of an optical fiber under load is inversely proportional to the quantity A and therefore water action on a loaded optical fiber may reduce its operational service life by a factor of 14-20.

Methods of Determining the Service Lives of Optical Fibers

The service life of an optical fiber τ under a constant load σ and having a maximum surface flaw of size a_{max} is determined by the expression

$$\tau = 2\,[(n-2)AY^n\sigma^n a_{max}^{(n-2)/2}]^{-1}, \qquad (5)$$

which may easily be derived from (1) and (3). In practice this formula cannot be used since we do not know the maximum flaw size along the entire optical fiber length. However if the optical fiber was would under a tensile force $\sigma_{пер}$, it is possible to estimate the maximum flaw size by using expression (2). Then the expression for the service life of such an optical fiber is written as

$$\tau = B\,\sigma_{пер}^{n-2}/\sigma_{раб}^{n}, \quad B = 2[(n-2)AY^2K_{кр}^{n-2}]^{-1}. \qquad (6)$$

As we see from (6) in order to determine the optical fiber service life we must know the value of the parameters n and A. Measurements indicate that $n = 21.5 \pm 1.5$. We did not succeed in determining the absolute value of the parameter A. We only established that an increase in crack growth rate in water is caused by an increase in this parameter.

After substituting our measured minimum value of $n = 20$ and the minimum value of $B = 10^{-3}$ $GPa^2 \cdot s$ from [13] into (6) we drafted graphs plotting the minimum optical fiber service life against the ratio of winding stress to operating stress.

Fig. 8 (solid line curves) reveal that in order to achieve a 25 year service life of the optical fibers it is necessary to achieve an operating stress that is 6-7 times lower than the winding stress.

Allowing the possibility of optical fiber failure with some low probability we may reduce the operational loading requirements. Taking advantage of the statistical nature of optical fiber strength and accounting for the fact that after winding the maximum flaw will have dimensions determined by expression (2) with some probability we may show that the optical fiber service life is determined by the expression [14]

$$\tau_F = \tau_{\text{пер}}(\sigma_{\text{пер}}/\sigma_{\text{раб}})^n \{[1 - \ln(1 - F)/(N_{\text{пер}}L)]^{(n-2)/m} - 1\}^{-1}_{,} \quad (7)$$

where $\tau_{\text{пер}}$ is the optical fiber time under loading during the winding process; F is the breaking probability of the optical fiber before the exhaustion of its service life; $N_{\text{пер}}$ is the average number of optical fiber breaks per unit of length in the winding process under a load of $\sigma_{\text{пер}}$; m is the slope of the graph of optical fiber unit strength plotted in Weibull coordinates near the load $\sigma_{\text{пер}}$; L is optical fiber length.

It follows from formula (7) that in this approach it is not necessary to know the absolute value of the constant B, although in this method estimates of optical fiber service life may be guaranteed only with a certain probability F. New optical fiber parameters $N_{\text{пер}}$, m and $\tau_{\text{пер}}$ also appear; these must be determined based on results from optical fiber winding under various loads.

Using typical data for our optical fibers under a winding force of 5 N: $\tau_{\text{пер}} = 2$ s; $N_{\text{пер}} = 5$ km^{-1}, $m = 10$, $L = 1$ km, $n = 20$ together with expression (7) Fig. 8 shows a family of graphs (dashed curves) representing the service life of optical fibers τ_F as a function of the ratio of $\sigma_{\text{пер}}/\sigma_{\text{раб}}$ for various optical fiber breaking probabilities. We may conclude from an examination of these graphs that less than 1% of optical fibers of kilometer length fail over a 25 year period if the ratio of the stress applied during winding to the operational stress is equal to three. If this ratio is equal to four then less than 0.01% of the optical fibers will fail over the 25 year period. When the winding tension is increased to 15 N the number of breaks $N_{\text{пер}}$ will increase to 10 km^{-1} while the value of the parameter m will drop to 1.5. After substituting these values into (7) we find that in the winding force range from 2-15 N the dependence of the probable optical fiber service life on the $\sigma_{\text{пер}}/\sigma_{\text{раб}}$ ratio remains virtually unchanged.

This water-activated flaw growth mechanism will be manifest, naturally, only on the optical fiber surface. We will now also examine flaws in the optical fiber bulk. Their growth is determined, as discussed above, by the thermal fluctuation mechanism investigated in the studies by Hilling, Bartenev, and Pukh for bulk vitreous samples [1, 2, 15]. The influence of this mechanism on optical fiber strength was discussed in a number of studies [16, 17]. The service life of optical fibers determined by the internal flaw growth rate is described in accordance with the thermal fluctuation theory, by the following formula [15]:

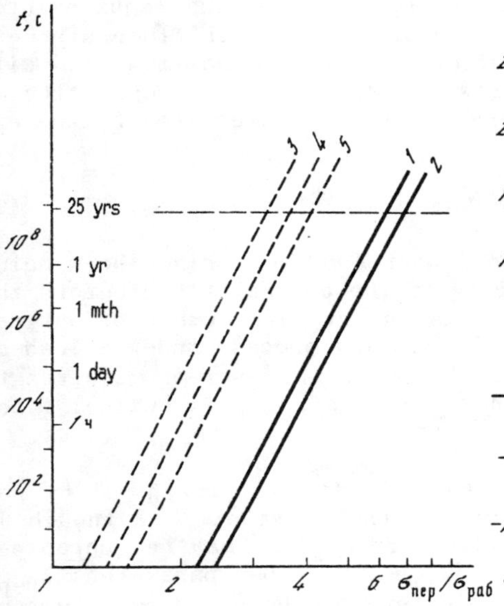

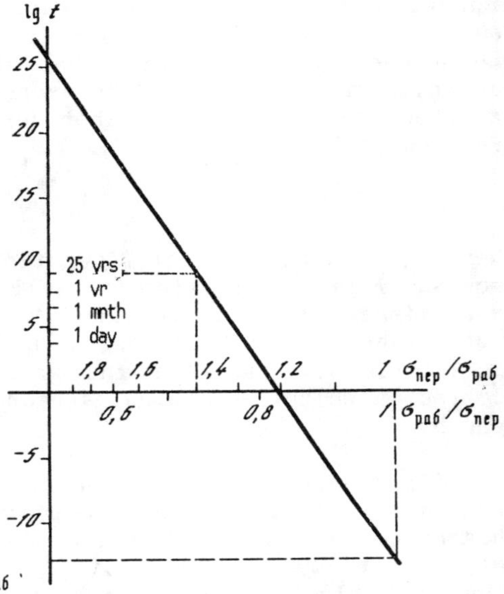

Fig. 8. Graphs plotting the minimum optical fiber service life (solid curves 1 and 2), winding force $P_{пер}$ = 5 (1) and 15 N (2) and the probable service life τ_F (dashed curves 3–5), winding force of 5 N and probability of optical fiber failure before exhaustion of service life $F = 10^{-2}$ (3), 10^{-3} (4), and 10^{-4} (5) as a function of the ratio of winding stress to operating stress

Fig. 9. The optical fiber service life with temperature-fluctuating flaw growth plotted as a function of the ratio of winding stress to operating stress

$$\tau_T = \tau_0 \exp [(U_0 - \gamma\sigma)/kT], \tag{8}$$

where $\tau_0 \approx 10^{-13}$ s is the critical time determined by the atomic thermal oscillation frequency; U_0 is the activation energy of the process and in the absence of water this is equal to the interatomic bond energy (for the SI-O bond, U_0 = 108 kcal/mol); k is Boltzmann's constant; γ is the constant determined by the properties of the material and the initial flaw size.

Expression (8) may be rewritten in the following form for an optical fiber that has been wound at a tensile stress level $\sigma_{пер}$

$$\tau_T = \tau_0 \exp U_0/kT\,(1 - \sigma_{раб}/\sigma_{пер}). \tag{9}$$

Fig. 9 plots the logarithm of the optical fiber service life as a function of the $\sigma_{пер}/\sigma_{раб}$ ratio (the temperature is taken as 300 K). As we see from this diagram the service life of the optical fibers drops below 25 years due to the thermal fluctuation flaw growth mechanism, i.e., this mechanism begins to become manifest if the ratio $\sigma_{пер}/\sigma_{раб}$ drops below 1.4. However at such a $\sigma_{пер}/\sigma_{раб}$ ratio the optical fiber service life determined by the growth of service flaws due to moisture will drop below a single day. Therefore the thermal fluctuation flaw growth mechanism will limit the service life of optical fibers only if their surface is not exposed to moisture.

In a more strict examination of the optical fiber service life estimates made necessary by the thermal fluctuation flaw mechanism it is necessary to account for the influence of thermal stresses that arise in the optical fiber due to the difference in the linear expansion coefficients of the core and coating materials. For standard optical fibers having a germanic oxide- or phosphorous monoxide-doped vitreous silica glass core and a reflective vitreous silica glass coating, the axial component of these stresses is added to the external tensile stresses in the optical fiber core, while in the reflective coating the thermal stresses weaken the external tensile stresses.

We will consider a specific example in order to determine the degree of influence of thermal stresses on the optical fiber service lives. When an optical fiber is wound at a force of 5 N an external tensile stress of 0.42 GPa is created in the fiber. In order to assure that this optical fiber will not fail after 25 years due to outer flaw growth, the operating stress cannot exceed 1/3 of the winding stress, i.e., 0.14 GPa. The temperature stresses on standard optical fibers with a numerical aperture of 0.2 is less than 0.07 GPa in the core and 0.02 GPa in the coating [8]. Therefore the actual stress in the core is 0.21 GPa and 0.12 GPa in the coating. If we compare these quantities to the actual winding stresses we find that in the core $\sigma_{пер}/\sigma_{раб} \approx 2.3$ while in the coating $\sigma_{пер}/\sigma_{раб} \approx 3.3$. Returning to Chapter 9 we may conclude that the service life of standard optical fibers determined by the thermal fluctuation flaw growth mechanism in this case is much greater than 25 years. As the winding stresses and, consequently, the operating stresses grow, the influence of thermal stresses on the optical fiber service lives will become even less significant. Thermal stresses evidently have a weak influence even in the failure of sealed surface optical fibers. Our multiple winding of optical fibers under tensile force has revealed that the internal flaws responsible for optical fiber failure are found either in the preform tube bulk material or on the boundary between the preform tube and the vitreous silica glass precipitated from the gas phase. Therefore in spite of the fact that the total stress present in the optical fiber core is less than the coating stress, the quantity and size of the flaws in the optical fiber protective coating will always be greater while the thermal fluctuation growth of precisely these flaws will determine the service life of sealed surface optical fibers.

Therefore we may draw the following conclusions from this discussion.

1. Systematic measurements of the mechanical strength of optical fibers and winding all fabricated optical fibers under tensile force in conjunction with an analysis of the causes of optical fiber failure have made it possible to significantly increase optical fiber strength and identify the ultimate strength of optical fibers fabricated by means of preform tubes of domestic manufacture.

2. It is possible to guarantee with a certain probability a 25 year service life of the optical fibers that have passed the winding process by selecting the proper ratio of winding force to operating tensile stress.

Bibliography

1. Khillig, U.B. "The causes of low mechanical strength and the ultimate strength of amorphous brittle solids" Prochnost' stekla [Mechanical strength of glass] Moscow: Mir, 1969, p. 68-120.

2. Bartenev, G.M. "Stroenie i mekhanicheskie svoystva neorganicheskikh stekol" [The structure and mechanical properties of inorganic glasses] Moscow: Stroyizdat, 1966, p. 216.

3. Kalish, D., Tariyal, B.K. "Static and dynamic fatigue of a polymer-coated fused silica optical fiber" J. AMER. CERAM. SOC., 1978, V. 61, p. 518-523.

4. Bogatyrev, V.A., Bubnov, M.M., Dianov, Ye.M., et al. "Investigation of the mechanical strength of optical fibers for optical communications systems" KVANTOVAYA ELEKTRON., 1981, V. 8, p. 844-852.

5. Mechlosky, J.J., Freiman, S.W., Morey, S.M. "Fractographic analysis of optical fibers" AMER. CERAM. SOC. BULL., 1977, V. 56, p. 1016-1017.

6. Becholsky, J.J., Rice, R.W., Freiman, S.W. "Prediction of fracture energy and flaw size in glasses from measurements of mirror size" J. AMER. CERAM. SOC., 1974, V. 57, p. 440-443.

7. Olshansky, R., Maurer, R.D. "Tensile strength and fatigue of optical fibers" J. APPL. PHYS., 1976, V. 47, p. 4497-4499.

8. Bogatyrev, V.A., Bubnov, M.M., Vechkanov, N.N., et al. "High-strength optical fibers fabricated by chemical deposition from the gas phase" KVANTOVAYA ELEKTRON., 1982, V. 9, No. 7, p. 1506-1509.

9. Justice, B., Runyan, M.A., Lafever, R.L. "Aging effects in optical fibers" AMER. CERAM. SOC. BULL., 1979, V. 58, p. 1126-1127.

10. Wang, T.T., Zupko, H.M. "Long-term mechanical behavior of optical fibers coated with a UV-curable epoxy acrylate" J. MATER. SCI., 1978, V. 13, p. 2241-2248.

11. "Bolokonno-opticheskaya svayz'" [Fiber-optic communications] Ed. by M.D. Khaues, D.V. Morgan, Moscow: Radio i svyaz', 1982, 179 p.

12. Helfinstine, J.D., Quan, F. "Optical fiber strength/fatigue experiments" OPT. AND LASER TECHN., 1982, V. 14, No. 3, p. 133-136.

13. Craig, S.P., Duncan, W.G., France, P.W. "The strength and fatigue of large flaws in silica optical fiber" 8th ECOC. Cannes, 1982, A VI-4, p. 205-209.

14. Mitsunaga, Y., Katsuyama, Y., Ishida, Y. "Reliability assurance for long-length optical fiber based on proof testing" ELECTRON. LETT. 1981, V. 17, p. 567-568.

15. Pukh, V.P. "Prochnost' i razrushenie stekla" [The mechanical strength and failure of glasses] Moscow: Nauka, 1973, p. 156.

16. Aleksandrov, I.V., Zhabotinskiy, M.Ye., Shushianov, O.Ye. "The reliability of vitreous silicone optical fibers" PIS'MA V ZhTF, 1984, V. 7, p. 949-952.

17. Aleksandrov, I.V., Zhabotinskiy, M.Ye., Shushiapov, O.Ye. "The mechanical reliability of optical fibers and possible methods of reliability improvement" ZhTF, 1984, V. 54, p. 1641-1662.

Temperature Resistant Fiber-Optic Modules

A.A. Abramov, M.M. Bubnov, N.N. Vechkanov, A.N. Gur'yanov,
A.S. Konov, V.N. Myakov, B.B. Troitskiy, A.G. Shchebunyaev

Abstract: The additional optical losses in fiber-optic modules with different primary and secondary polymer coatings are investigated. It is discovered that the increase in optical losses in the low temperature range may be attributed primarily to crystallization of the silicone elastomer in the primary coatings. Fiber-optic modules with a primary coating consisting of a specially-developed noncrystallizing PFS-2 silicone elastomer were fabricated; these modules have an optical loss increment of less than 1 dB/km for temperature variations from -120 to 200°C.

Introduction

In order to achieve reliable operation of fiber-optic communications systems it is necessary to eliminate climatic factors, primarily temperature influences on optical losses and other transmission characteristics of such systems. Optical losses of optical fibers fabricated from vitreous silica glass remain virtually unchanged over a broad (-150 to 150°C) temperature range [1, 2]. However when polymer jacketings are applied to the optical fibers the attenuation of the transmitted signal becomes highly dependent on variations in external temperature. A significant growth of optical losses has been observed in optical fibers in the low temperature range. This is due to the fact that as a result of the significant difference in the coefficients of linear expansion of the vitreous silica glass and the polymer materials comprising the protective jacketings the thermal shrinkage of such jacketings in the cooling process causes optical fiber microbends which then serve to increase optical losses [3]. Since the application of polymer coatings to optical fibers are necessary to preserve strength the problem of developing a fiber-optic module design (an optical fiber with one or several polymer jacketings) arises as well as the problem of selecting polymers that will minimize additional losses in the optical fiber across the entire operating temperature range.

Extensive foreign research (see, for example, [3-7]) devoted to the increase in optical fiber losses with changes in external temperature have revealed that the level of such additional losses is determined by a wide range of factors including the core diameter and the external optical fiber diameter, its numerical aperture, the Young's modulii and the coefficients of linear expansion of the polymer jacketing materials and the thickness of these layers. Moreover the additional losses in optical fibers from temperature variations depend on how densely the polymer jacketing is applied or whether lapping is used [8] as well as the application conditions of these jacketings [8, 9]. The present study investigates optical fibers employing protective jacketings of various polymer materials of domestic manufacture in order to develop fiber-optic modules that may function over a broad temperature range.

1. Fiber-Optic Module Structure and the Technique for Measuring Additional Optical Fiber Losses

The most widely used fiber-optic module design is one employing two polymer jacketings [7]. The primary characteristics of such a module include E_i, γ_i, D_i, and S_i: the Young's modulus, the coefficient of linear expansion, the diameter, and the cross-sectional area, respectively, with the index $i = 0$ referring to the optical fiber itself, $i = 1$ to the primary jacketing and $i = 2$ to the secondary jacketing. Soft polymers (silicone elastomers and modified polyacrylates) with a small Young's modulus (from 10^{-3} to $2 \cdot 10^{-2}$ GN/m^2), while more rigid polymers are used for the secondary jacketing (polyamides, fluoroplasts (Teflons), polypropylenes, and polyethylenes) with a Young's modulus from 0.1-5 GN/m^2 [6]. Since the soft buffer jacketing effectively protects the optical fiber from lateral loading, such a design of the fiber-optic module provides a low increment of optical losses both during the application of the polymer jacketings and in the manufacture of optical cable. Characteristics of certain materials employed here are given in Table 1.

When optical fibers are cooled their polymer jacketings shrink in both the radial and axial (longitudinal) directions therefore causing microbends in the optical fiber. Research conducted in study [3] demonstrated that the primary cause of increasing losses at low temperature derive from the shrinkage of polymer jacketings in the axial direction. If some method is used to reduce longitudinal optical fiber compression from cooling (for example by lapping the optical fiber onto a large diameter silica rod), the level of additional losses may be significantly lower than from the free coiling of the optical fiber. Radial compression evidently makes a significant contribution to the optical loss increment from optical fiber cooling in the case of a nonconcentric configuration of the optical fiber within the polymer jacketings.

Following study [6] we will consider the forces that arise in the optical modulus from a change in temperature. forces parallel to the optical fiber axis arise in each polymer layer and in the vitreous silica glass optical fiber from cooling:

$$F_i = \frac{L - L_i(T)}{L_i(T)} S_i E_i, \tag{1}$$

where L is optical fiber length at temperature T; $L_i(T)$ is the equilibrium length of the i-layer at temperature T. Since

$$L_i(T) = L_i(20^\circ)(1 + \gamma_i \Delta T), \tag{2}$$

where $\Delta T = T - 20^\circ$ and accounting for the fact that $\gamma_i \Delta T \ll 1$, we obtain

$$F_i = (\epsilon - \gamma_i \Delta T) S_i E_i. \tag{3}$$

Here $\varepsilon = [L - L(20^\circ)]/L(20^\circ)$.

Table 1

Material characteristics (at $T = 20^\circ$C)

Material	Young's modulus, E, GN/m^2	Coefficient of linear expansion γ, deg^{-1}
Vitreous silica glass	73	$3.4 \cdot 10^{-7}$
Silicone elastomer	$\sim 1.2 \cdot 10^{-3}$	$\sim 10^{-4}$
Polyamide P-610	1.5–1.7	$\sim 10^{-4}$
Fluoroplast (Teflon) F-2M	0.8–1.2	$\sim 10^{-4}$
Fluoroplast (Teflon) F-4MB	0.50-0.65	$\sim 10^{-4}$

If the fiber-optic module is in thermal equilibrium and there is no slippage between the optical fiber and the polymer jacketings, then

$$\sum_i F_i = \epsilon \sum_i S_i E_i - \sum_i S_i E_i \gamma_i \Delta T = 0. \tag{4}$$

Then the relative elongation of the fiber-optic module is equal to

$$\epsilon = \frac{\sum_i S_i E_i \gamma_i}{\sum_i S_i E_i} \Delta T = \gamma \Delta T, \tag{5}$$

where

$$\gamma = \sum_i S_i E_i \gamma_i / \sum_i S_i E_i. \tag{6}$$

The coefficient of linear expansion of the fiber-optic modulus γ is its most important characteristic making it possible to estimate the degree of sensitivity of the optical losses of the modulus to temperature changes. For example, as established in study 9 for modules with a polyamide protective jacketing the additional losses from cooling are

$$\Delta a \sim \gamma^3. \tag{7}$$

After substituting the values of E_i, γ_i, and S_i for the specific fiber-optic module designs into (6) we may estimate the relative contribution of the various jacketings to the total coefficient of linear expansion γ. In this case we must account for the fact that the values of E_i and γ_i of the polymer materials may change by several orders of magnitude with a change in temperature [10]. At near-room temperatures (for modules with D = 0.125 mm, D_1 = 0.25 mm, and D_2 = 0.5 mm) the following relations are normally satisfied: $S_2E_2\gamma_2 \gg S_1E_1\gamma_1$, $S_0E_0 \gg S_1E_1$, $S_0E_0 > S_2E_2$, $S_2E_2\gamma_2 > S_0E_0\gamma_0$. In this case formula (6) takes the following approximate form:

$$\gamma \cong S_2E_2\gamma_2/S_0E_0, \tag{8}$$

i.e., the smaller the cross-section (and therefore the thickness) of the secondary jacketing, the less sensitive the fiber-optic modules to changes in temperature.

In investigating the additional optical losses that arise from the cooling of fiber-optic modules with different primary and secondary polymer jacketings we employed optical fibers fabricated by the technology developed at the Institute of Chemistry of the Academy of Sciences and the Institute of General Physics of the Academy of Sciences by chemical deposition from the gas phase. These optical fibers had a step refractive index and standard core dimensions (50 μm) and external diameter dimensions (125 μm). The numerical aperture of the optical fibers varied between 0.2 and 0.23.

The primary silicone 60-65 μm coating was applied to the optical fibers in the drawing process. The FV-2 high-modulus polyacrylate lacquer, polyvinyl chlorine plastisol, P-610 polyamide as well as the F-2M and F-4MB fluoroplasts were used as the secondary coating materials. A 5-10 μm polyacrylate coating and a polyvinyl chlorine coating approximately 120 μm in thickness were applied to the optical fibers in their drawing process. The 120-310 μm polyamide and fluoroplast layers were applied after drawing while the P-610 polyamide was applied by means of a melting crucible while the fluoroplasts were applied by extrusion. Application of the secondary coatings did not cause a significant increase in optical fiber optical losses. Table 2 gives the designs of several of the optical fiber modules we fabricated and tested.

An assembly analogous to that used in study [11] at 0.85 μm was employed to measure the additional optical losses occurring in optical

Table 2

Test fiber-optic module designs

Module number	Jacketing materials	Jacketing thickness, μm
1	SIEL 159/167	65
2	SIEL 159/167-FV-2	65-5
3	PS-1-FV-2	65-5
4	PFS-2-FV-2	65-5
5	FV-2	5
6	SIEL 159/167-PVKh	12
7	SIEL 159/167-FV-2-F-2M	65-5-230
8	SIEL 159/167-P-610	65-125
9	P-610	375
10	PS-1-FV-2-F-2M	65-5-120
11	PFS-2-FV-2-F-4MB	65-5-120
12	SIEL 159/167-FV-2-F-4MB	65-5-120

fibers under cooling. The absolute measurement accuracy was better than 0.01 dB. In developing this assembly special attention was focused on its operating stability over a long measurement period (10-12 hours). In order to monitor the stability of the light source and the operation of the measurement instrumentation the optical scheme incorporated a reference channel.

We established in the development of the measurement technique that the level of additional losses is highly dependent on the optical fiber lapping technique (tensile force, number of layers, lapping regularity), the degree of surface roughness of the rod onto which the optical fiber is wound and on the material comprising this rod. Analogous effects were also discovered in study [3]. In this study, for example, when the optical fiber is wound onto a vitreous silica glass rod there was no increase in optical losses discovered in the low temperature range, since such a rod eliminated shrinkage of the polymer coatings from cooling and, consequently, significantly reduced optical fiber microbending. In order to eliminate the influence of the winding and the rod on measurement results we developed a rod design with a variable internal diameter. The initial diameter of the duralumin rod was 180 mm. After winding of the optical fiber onto the rod we reduced the diameter of the rod somewhat prior to the measurement process, thereby allowing free optical fiber lapping.

The test optical fibers were cooled in a nitrogen chamber at a mean rate of ~ 1 deg/min [12]. A thin aluminum foil layer was applied between the optical fiber layers to assure uniform cooling and as a result the temperature did not vary by more than 2-3°C across the entire temperature (from -150 to 200°C).

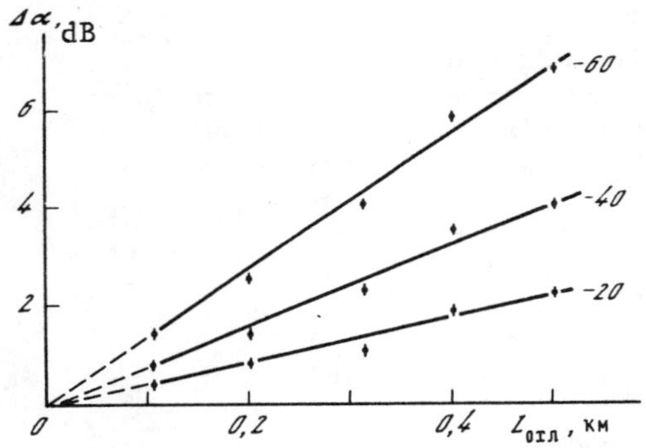

Fig. 1. Additional optical losses $\Delta\alpha$ plotted as a function of the length of the cooled optical fiber section $L_{охл}$
The numbers to the right of the curves indicate temperature in °C

Normally the test optical fibers were 500 m in length and were excited by light with a near-fixed aperture. In order to eliminate Lossy modes the polymer jacketings near the optical fiber ends were removed and the bare segments were immersed. The first 200 m of the optical fiber were not cooled and were used as the excitation optical fiber in which the stationary modal distribution was established for the given optical fiber segment. The dependence of additional losses on the length of the cooled optical fiber section with this measurement technique was linear (Fig. 1) and therefore the measurement results for the additional losses were converted into decibels per kilometer.

2. The Influence of Primary Coatings on Additional Losses in Optical Fibers at Low Temperatures

In order to determine the relative contribution of the primary and secondary polymer jacketings to the additional losses of the fiber-optic modules at low temperatures we initiated our investigations by measuring the additional low-temperature losses in optical fibers having only primary silicone coatings. As indicated by our investigations [12] under short term exposure to low temperatures to-75°C there was no increase in optical losses in these optical fibers. However further cooling of the optical fiber below -80°C caused a sharp increase in optical losses (Fig. 2, optical fiber 1). The drop in optical losses from heating at a temperature of approximately-40°C, i.e., the temperature dependence of the additional losses revealed a hysteresis nature.

Such a temperature dependence of the additional losses may be traced to the fact that in the -120 to -40°C range the silicone elastomer used in the primary jacketing crystallizes, i.e., it undergoes a first order phase transition accompanied by a sharp change in the mechanical and physical properties of the polymer: the elastomer becomes more rigid, its elasticity is degraded, the module expands, its volume drops, and therefore its density increases, which causes optical fiber microbending.

The rate of crystallization of the silicone elastomer is temperature dependent [13]. The crystallization rate is at a peak in the center of the crystallization range ($T_{кр.макс} \simeq -80°C$) and vanishes at its edges. The boundaries on the crystallization range are determined, on the one hand, by the vitrification temperature of the elastomer ($T_{ст}$) and, on the other, by its melting point ($T_{пл}$).

The melting point may change as a function of the crystallization conditions, although in our investigations at a relatively slow (~1 deg/min) change in temperature $T_{пл}$ it was accurately reproduced (the temperature history of the elastomer and the crystallization conditions had virtually no influence on the reverse progression of the temperature dependence).

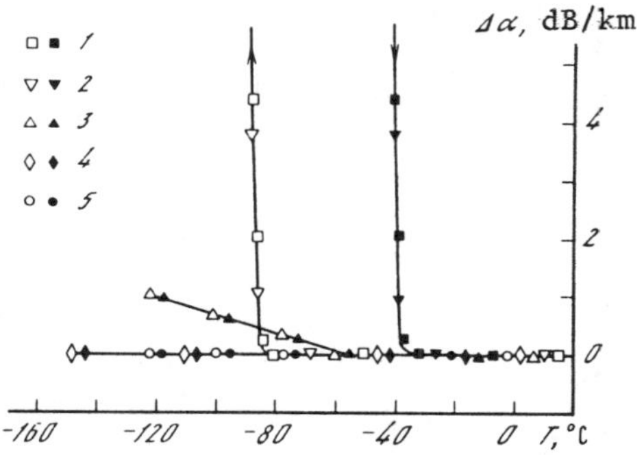

Fig. 2. Additional optical losses in optical fibers (1-5) with different primary coatings
The open symbols represent cooling while the solid symbols represent heating

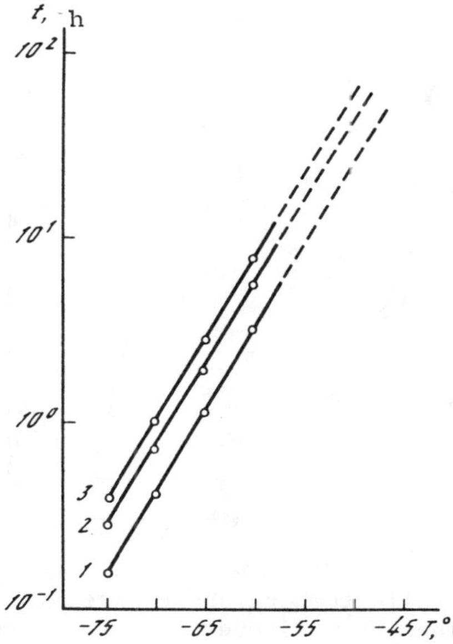

Fig. 3. The temperature dependence of the loss rise time t to specific levels in optical fiber 1
1 - to 1 dB/km; 2 - to 5, 3 - to 10

The dependence of the crystallization rate of the SIEL 159-167 silicone elastomer on temperature is illustrated by Fig. 3 in which the temperature dependencies of the loss rise time in the optical fiber to specific levels (1.5 and 10 dB/km) are shown. It is clear from the graphs in Fig. 3 that this elastomer crystallizes at $T = -60°C$ sufficiently slowly and therefore in order to fix a rise in losses of 1 dB/km in an optical fiber in an SIEL 159-167 jacketing it is necessary to expose this optical fiber to -60°C for at least 4 hours.

In actual operating conditions the optical fibers may be exposed to low temperatures for a significantly longer period and therefore the lower functional limit of an optical fiber with a crystallizing silicone elastomer jacketing should be the equilibrium melting point of its crystalline phase, i.e., the temperature above which the crystalline phase cannot exist thermodynamically.

The SIEL 159-167 elastomer vitrifies at near -120°C and it is therefore difficult to experimentally observe the influence of vitrification on optical optical fiber losses due to the high additional loss level caused by elastomer crystallization. However by very rapid (~10-15 deg/min) cooling of the optical fiber to -120∘C it is possible to "jump over" the crystallization range without noticeable formation of the crystalline phase. The temperature dependence of the additional losses from vitrification, unlike crystallization, manifests no hysteresis nature (Fig. 4, the second "cooling-heating" cycle).

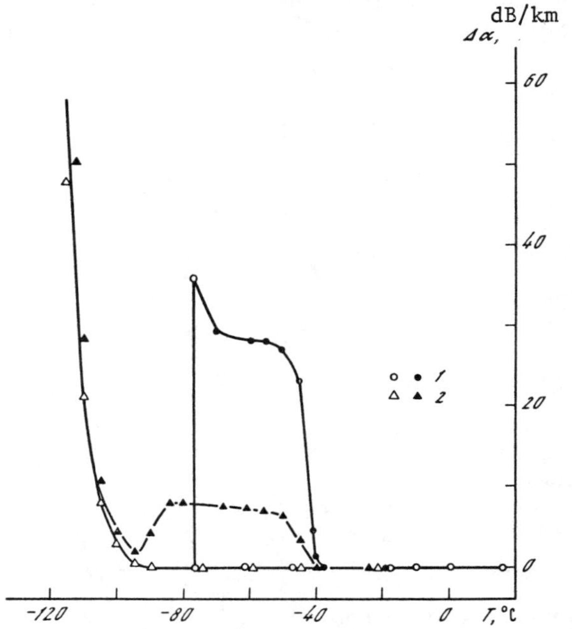

Fig. 4. The temperature dependence of additional optical losses under cyclic temperature changes for optical fiber 1
The open figures represent cooling and the solid figures represent heating. In cycle 1 the optical fiber was exposed to a temperature of -77∘C for 30 minimum at a heating and cooling rate of ~1 deg/min while in cycle 2 the cooling rate was 10-15 deg/min with a heating rate of ~1 deg/min

Additional functional groups may be introduced to suppress crystallization of the silicone elastomers. We developed such noncrystallizing elastomers. Crystallization in these elastomers was suppressed by introducing either additional (compared to the SIEL 159-167 elastomer) hydride and vinyl groups (elastomer B-1) [14] or phenyl groups (elastomers PS-1, PFS-2) [12]. Results from measurements of the additional losses in optical fibers 3 and 4 (see Fig. 2) with primary noncrystallizing elastomer coatings revealed that no hysteresis of the optical losses was observed in these optical fibers, while the additional loss level at temperatures below -40°C was much less than in optical fiber 1 with a crystallizing jacketing. We should note that optical fibers 3 and 4 in addition to the silicone primary jacketing also had a thin (5-10 μm) protective FV-2 polyacrylate lacquer coating. Our research has indicated that such a thin coating makes no contribution to additional low-temperature losses (for example, optical fibers 1 and 2), although it does provide efficient failure protection to the primary coating. The insignificant increase in losses in optical fiber 3 at temperatures below -60°C could possibility be related to the vitrification of the PS-1 elastomer or the unavoidable shrinkage of the elastomer under cooling. We did not observe additional losses from cooling to -120°C in optical fiber 5 having only a single thin lacquer coating 5-10 μm in thickness, although optical fibers with such a coating do not have sufficient mechanical strength [15] and therefore are difficult to employ in optical communications systems.

Therefore our investigations have revealed that crystallization of the silicone elastomers in the primary jacketings significantly narrows the operational temperature range of the optical fibers. Therefore it is necessary to use noncrystallizing elastomers with a minimum vitrification temperature as the primary jacketings in optical fibers. In other words the necessary condition to guarantee low additional losses from optical fiber cooling is the application of polymers with no phase transitions across the entire operational temperature range.

3. The Influence of Secondary Coatings on Additional Optical Losses in Optical Fibers at Low Temperatures

As noted earlier in section 1 the most popular fiber-optic modules today are those employing two polymer jacketings: a soft primary jacketing and a hard secondary jacketing. The secondary jacketing is used primarily to protect the optical fiber from lateral compressive loads. The thicker this jacketing the more effectively the optical fiber is protected [16]. At the same time expression (7) reveals that the increase in secondary jacketing thickness causes an increase in additional losses from optical fiber cooling. An analysis of the fiber-optic modules from study [7] has revealed that in order to both maintain a low increment of optical losses at low temperatures and a low increment of optical losses from lateral compression it is necessary for the following conditions to be satisfied: $0.4 \leqslant D_1/D_2 \leqslant 0.6$ and $0.4\ \text{mm} \leqslant D_2 \leqslant 0.6\ \text{mm}$. Fiber-optic modules 6-8 and 10-12 were fabricated with these requirements in mind (see Table 2). Results from measurements of the additional losses in these modules as a function of temperature are given in Fig. 5 which clearly indicates that when the SIEL 159-167 crystallizing elastomer is used as the primary coating the temperature dependence of the optical losses manifests a hysteresis nature, and the greater the Young's modulus of the material in the secondary jacketing, the earlier optical loss growth begins [17]. It is possible that in this case pressure on the silicone elastomer of the secondary jacketing stimulates its crystallization process [13]. Under heating the optical losses begin to drop at approximately the same temperature as the melting point of the crystalline phase in the elastomer.

It is possible to significantly reduce additional losses in optical fibers with a dual-layer coating by using noncrystallizing elastomers (PS-1, PFS-2) as the primary coating material. For such optical fibers additional optical losses are an unambiguous function of temperature, and no hysteresis of optical losses is observed (see Fig. 5b, modules 10 and 11). The higher additional loss level for module 10 may be attributed to the difference in the mechanical characteristics of fluoroplasts F-2M and F-4MB as well as the difference in the mechanical properties of the primary coating materials used for these modules (PS-1 and PFS-2). We were not able to fix an increase in optical losses for module 11 even by cooling this module below the vitrification temperature of the PFS-2 elastomer (−105°).

With an increase in the thickness (cross-section) of the secondary jacketing the coefficient of linear expansion of the module γ grows in accordance with (6) and (7) and optical losses also increase (Fig. 6). For our optical fibers with a polyamide secondary jacketing the additional losses are proportional to the cube of the coefficient of linear expansion of the module. A similar relation was established for optical fibers with a secondary polyamide jacketing in study [3]. This relation takes the following form for optical fibers with F-2M fluoroplast jacketing

$$\Delta a \sim \gamma^2. \qquad (9)$$

Such a weaker dependence of additional losses on the coefficient of linear expansion γ may clearly be traced to the fact that, first, the fluoroplast coatings were applied to the optical fibers at a very low compression level and, second, as indicated in Table 2, optical modules with fluoroplast coatings had one additional intermediate lacquer jacketing. This jacketing reduces friction between the silicone and fluoroplast coatings and therefore reduces axial compression of the optical fiber from the thermal shrinkage of polymer jacketings and as a result reduces the level of additional optical losses of the optical fiber in the low temperature range [18].

We also investigated optical fibers with a single hard protective jacketing (see Fig. 5a, module 9). It turned out that it is possible to achieve a sufficiently low additional loss level in such optical fibers due to the limited capacity of a vitreous silican optical fiber with such a jacketing to bend. However the additional loss level in optical fibers with the hard jacketing is highly dependent on the nonconcentricity of the jacketing as well as its homogeneity.

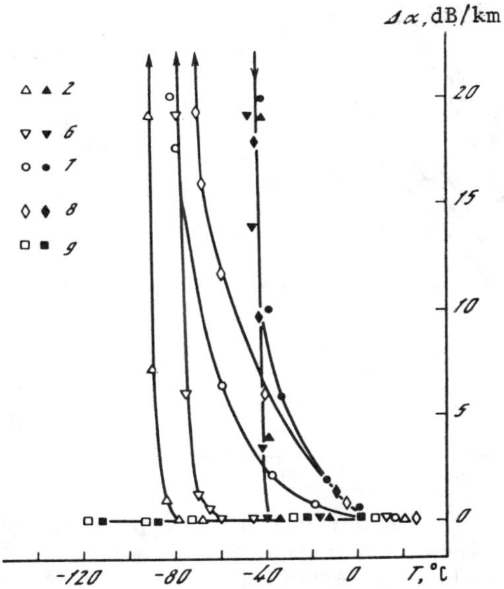

a

Fig. 5. Additional optical losses in modules with an SIEL 159-167 silicone elastomer primary coating and various secondary coatings (a) and the temperature dependencies of the additional optical losses for modules with PS-1 and PFS-2 noncrystallizing elastomer primary coatings (b)
The open symbols represent cooling and the solid symbols represent heating; 2, 6-12-fiber-optic modules

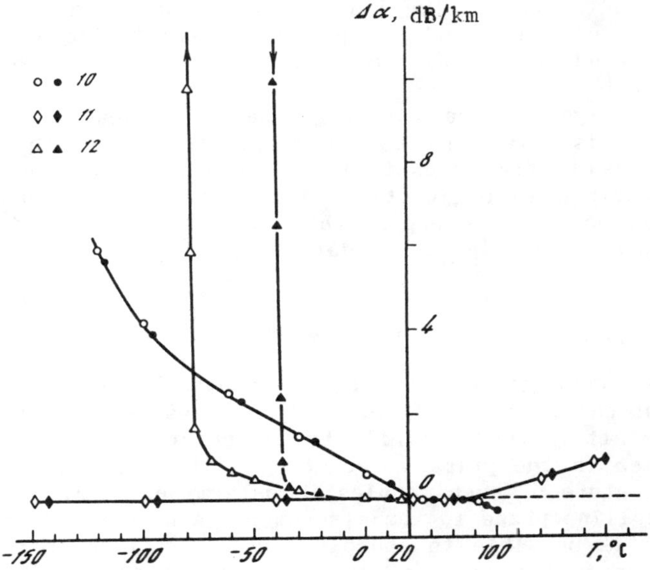

b

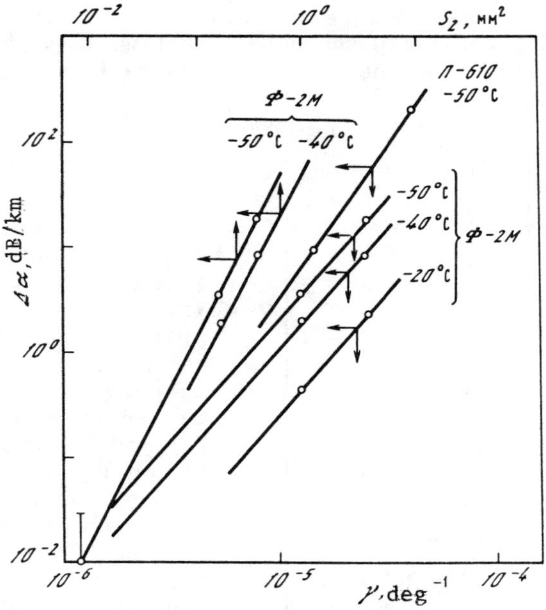

Fig. 6. The additional low-temperature optical losses plotted as a function of the coefficient of linear expansion of the module γ and the cross-sectional area of the secondary jacketing S_2

Moreover optical fibers without the soft reinforcing jacketing are very sensitive to compressive loading [4]. Therefore in our view the fiber-optic module design with a soft noncrystallizing primary coating and a hard secondary coating are the most universally advantageous.

In the high temperature range the functional boundary of the optical fibers is determined by the temperature strength of the polymers in the protective jacketings. Employing polysiloxanes as the primary jacketings in conjunction with fluoroplast secondary jacketings makes it possible to expand the temperature range of the optical fibers to +200∘C (see Fig. 5b, module 11).

Conclusion

Results from these investigations have revealed that the increase in optical losses in fiber-optic modules in the low temperature range may be attributed primarily to the crystallization of the silicone elastomer in the primary jacketing. Therefore the lower operational temperature of such modules is determined by the melting point of the crystalline phase in the elastomer. A noncrystallizing primary coating made it possible to sharply reduce the loss increment and to broaden the operating range of the optical fibers to -150∘C. Fiber-optic modules with a primary PFS-2 noncrystallizing silicone elastomer (developed by the authors) were fabricated having an optical loss increment of less than 1 dB/km over a temperature range of -120 to 200°C.

Bibliography

1. Yeung, W.F., Johnson, A.R. "Effect of Temperature on Optical Fiber Transmission" APPL. OPT., 1978, V. 17, p. 3703-3705.

2. Andreev, A.Ts., Gur'yanov, A.N., Konov, A.S., et al. "Investigation of the temperature dependence of optical losses in low-loss optical fibers" KVANTOVAYA ELEKTRON., 1980, V. 7, p. 2210-2213.

3. Sugaware, Y., Kobayashi, T., Tanaka, M., et al. "Attenuation increase mechanism of jacketed and cabled fibers at low temperature" Proc. 5th Europ. Conf. Opt. Commun., Amsterdam, 1979, p. 12.4-1-12.4-4.

4. Naruse, T., Sugaware, Y., Masuno, K. "Nylon jacketed optical fiber with silicone buffer layer" ELECTRON. LETT., 1977, V. 13, p. 153-154.

5. Inada, K., Sugawara, Y., Kobayashi, T. et al. "Cabling of ultra-low-loss optical fibers" Proc. 4th Europ. Conf. Opt. Commun., Geneva, 1978, p. 200-203.

6. Hanson, E.G. "Origin of temperature dependence of microbending attenuation in optical fiber cables" FIBER and INTEGRAT. OPT. 1980, V. 3, p. 113-148.

7. Masuno, K., Ishihara, K. "Optimum design of coated optical fiber considering excess loss at low temperatures" J. OPT. COMMUN., 1982, V. 3, p. 142-145.

8. Rokinohe, M., Shintani, T., Yajima, M. "Stability of transmission properties of optical fiber cables" Proc. 2nd Europ. Conf. Opt. Fiber Commun., Paris, 1976, p. 183-188.

9. Suzuki, H., Osanai, H. "Effect of secondary coating technique of low-loss optical fibers on transmission loss" Organic Coatings and Plastics Chemistry, Honolulu: Amer. Chem. Soc., 1979, p. 211-216 (CHEM. SOC. JAP. CHEM. CONGR.; V. 40).

10. Kimura, T., Yamakawa, S. "New UV-curable primary coating material for optical fiber" ELECTRON. LETT. 1984, V. 20, p. 201-202.

11. Belov, A.V., Dianov, Ye.M., "Investigation of total loss dependence in optical fibers on radiation injection angle" KVANTOVAYA ELEKTRON., 1979, V. 6, p. 404-407.

12. Abramov, A.A., Borkina, G.Yu., Bubnov, M.M., et al. "The influence of primary polymer coatings on the optical losses of optical fibers at low temperatures" KVANTOVAYA ELEKTRON., 1985, V. 12, p. 839-841.

13. Bukhina, M.F. "Tekhnicheskaya fizika elastomerov" [Technical physics of elastomers] Moscow: Chemiya, 1984, p. 171.

14. Andreev, A.Ts., Borkina, G.Yu., Bubnov, M.M., et al. "The influence of gamma-radiation on the temperature dependence of "vitreous silica glass-polymer" optical fibers" KVANTOVAYA ELEKTRON., 1981, V. 8, p. 1816-1817.

15. Huber, P., Guttmann, J. "Reproducibility of optical and mechanical properties of fibers prepared by the modified chemical deposition process" PHYS. AND CHEM. GLASSES. 1980. V. 21, p. 43-46.

16. Katsuyama, Y., Mitsunaga, Y., Tanaka, C., et al. "Optical loss stability of a fiber cable under lateral force and optimum design of the coated fiber structure" APPL. OPT., 1982, V. 21, p. 1337-1341.

17. Abramov, A., Borkina, G., Bubnov, M., et al. "Frost-resistant coated optical fibers" ELECTRON. LETT, 1985, V. 21, p. 805-807.

18. Okagawa, S., Nishinura, M., Mizota, Y., et al. "A new UV-resin/-nylon tightly jacketed optical fiber having excellent low-temperature properties" Proc. 5th Intern. Conf. Integr. Optics and Opt. Fiber Commun. and 11th Europ. Conf. Opt. Commun., Venezia, 1985, p. 519-522.

Optical Properties of Germanium Silicate Glass for Low-Loss Optical Fibers

V.M. Mashinskiy

Abstract: The material dispersion of germanium silicate glass in the form of MCVD-preforms and optical fibers is investigated and quantitative data are derived on the spatial distributions of doping germanium and boron additives and hydroxyl impurities. The influence of industrial processes, particularly the optical fiber drawing process, on these parameters is determined. The interrelationship between the absorption and luminescence bands and the formational conditions of germanium centers are established based on an investigation of the spectral luminescence characteristics of the germanium centers in germanium silicate glass and on their radial distributions. The absence of preferential association of chlorine with both germanium and the germanium centers is demonstrated.

Introduction

Over the last 20 years from the initial ideas for optical fiber applications in communications links to the present, fiber-optics has expanded into an extensive branch of science and technology, fiber-optic communications links are being used on a broader basis in practice and optical fibers are finding increasingly new applications. All these successes have been based on the extraordinary properties of SiO_2 vitreous silica glass which is used as the basis for virtually all optical fibers. Other materials whose optical properties may be potentially better than those of SiO_2 are at the present under intense scientific investigations or have been studied previously and so far present no competition to vitreous silica glass.

A variety of doping additives (GeO_2, B_2O_3, P_2O_5, TiO_2) are used to generate the waveguide structure; germanium dioxide predominates from this group in a number of physical properties. Today both scientific and engineering interest is shifting towards single-mode optical fibers whose manufacture widely employs doping of vitreous silica glass by fluorine. Nonetheless germanium silicate glass (germanium dioxide doped vitreous silica glass) is one of the primary low loss optical fiber materials. The advantages of such glass include, first, that at low GeO_2 concentrations (2 mol%) optical fiber losses are virtually equal to the natural vitreous silica glass loss level of 0.2 dB/km at λ = 1.55 μm [1] and, second, the fact that as the GeO_2 concentration rises to 20 mol% changes in the technologically important parameters of the glass (viscosity, coefficient of thermal expansion) are relatively insignificant which makes it impossible in turn to obtain high-numerical aperture optical fibers (*NA* = 0.29) while preserving the low loss factor (less than 1 dB/km at λ = 1.55 μm) [2].

The present study gives results from select investigations of germanium silicate glass principally in the form of preforms and optical fibers focusing on the dispersion properties of optical fibers, impurity absorption in the optical fibers and the properties of germanium silicate glass flaws.

The Material Dispersion of Germanium Silicate Glass. A Comparison of Preform and Fog Optical Fiber

The material dispersion *M* causes temporal broadening of the pulses $\Delta\tau$ with a final spectral width $\Delta\lambda$ which establishes the upper frequency band limit for the given radiator

$$\Delta\tau = -M(\lambda)\,\Delta\lambda = -\frac{\lambda}{c}\frac{d^2 n(\lambda)}{d\lambda^2}\Delta\lambda, \tag{1}$$

where $n(\lambda)$ is the refractive index of the medium; λ and c are wavelength and the speed of light in the vacuum. A point λ_0 always exists in the transparency region of the medium where $M(\lambda_0) = 0$ and, consequently, the broadest bandwidth of the fiber-optic communications link will be achieved in the vicinity of this point.

There exist several methods of measuring the material dispersion in optical fibers and bulk glass samples [3-6]. If the early methods are of greater practical value, the latter techniques have greater accuracy and characterize the properties of the material in pure form. At the same time using these results for optical fibers makes it necessary to account for the inhomogeneity of the optical fiber core material as well as possible changes from drawing. Therefore it is fundamentally important to carry out measurements of the material dispersion in the preform itself and to compare to results obtained in an optical fiber drawn from this same preform [7].

Measurements of the dispersion of $n(\lambda)$ were carried out on a preform fabricated by the MCVD process with a step refractive index and a mean GeO_2 concentration of app. 11 mol% in the core. The measurements were performed using the least deviation method over a broad spectral range of 0.8-2.4 μm. The accuracy of the assembly was $\pm 1 \cdot 10^{-4}$. Measurements were also carried out for $n(\lambda)$ in various types of vitreous silica glass. The measurement results were approximated by the trinomial Sellmeier formula

$$n^2(\lambda) - 1 = \sum_{i=1}^{3} \frac{a_i \lambda^2}{\lambda^2 - b_i} \tag{2}$$

and were used to calculate the material dispersion $M(\lambda)$ by (1). The coefficients of the Sellmeier formula, the value of λ_0 and the standard error are given in Table 1.[1] It is clear that the material dispersion for the core shifts towards longer wavelengths compared to vitreous silica glass and vanishes at $\lambda_0 = 1.303$ μm. These data virtually coincide with the results from study [5] which employed a set of selective emission sources while the measurements were carried out in uniform bulk glass samples.

Table 1

Coefficients of the Sellmeier formula, the standard error and the wavelength of zero material dispersion

material	a_1	b_1	a_2	b_2	a_3	b_3	σ, 10^{-4}	λ_0, мкм
$0.89 SiO_2 - 0{,}11 GeO_2$ (core no. 57)	0.657716	0.0132053	0.493315	0.0010052	0.856899	98.06363	1.89	1.303±0.015
SiO_2 (core no. 57)	0.660573	0.0030566	0.450205	0.0134082	0.926668	98.02213	1.64	1.242±0.015
KSG	0.653953	0.0143108	0.450441	−0.002332	0.877785	98.06261	0.27	1.267±0.002
KV	0.654024	−0.003291	0.449927	0.0233887	0.861180	98.11413	0.31	1.285±0.002
KV70 (non aqueous)	0.654083	0.0159573	0.450626	−0.0053221	0.852863	98.91244	0.26	1.275±0.002

The method from study [6] was used to determine the material dispersion in an optical fiber drawn from the test preform; this principle involved the direct measurement of the spectral dependence of the squared numerical aperture of the optical fiber:

$$NA^2(\lambda) = n_1^2(\lambda) - n_2^2(\lambda), \tag{3}$$

where n_1 and n_2 are the refractive indices of the core and the cladding of the optical fiber, respectively, and $M(\lambda)$ was calculated based on results from measurements of $NA^2(\lambda)$ and given values of n_2 obtained by an independent method.

---

[1] The author wishes to acknowledge the contribution of A.B. Grudinin and L.A. Frenkel' who pointed out that the data given in the table in article 7 are incorrect due to an arithmetical error in the approximation of the experimental data.

The relative spectral distribution $NA^2(\lambda)$ was found from the ratio of intensities of emission exiting an optical fiber approximately 0.5 m in length when the optical fiber is excited by a beam with an aperture that is deliberately greater than the critical value in the first measurement and deliberately less than the critical value in the second measurement. The measured aperture of the helium-neon laser beam exiting the optical fiber was used to fix to the absolute values of NA (λ = 0.6328) when exciting all optical fiber modes in the optical fiber. As a result the measured numerical aperture of the optical fiber was 2-4% below that of the preform (in an overall measurement range of 0.8-1.6 μm). The cause of this discrepancy may be found in the reduction in the average germanium concentration in the core due to its diffusion to the cladding (which will be examined below) as well as the influence of thermoelastic stresses which, through the photoelastic effect, reduces the numerical aperture of the optical fiber by approximately 2% according to our estimates. If we account for the fact that the accuracy of the $NA(\lambda)$ measurements was approximately 5%, we may assume that the numerical aperture and, consequently, the material dispersion of the germanium silicate glass do not change in the optical fiber drawing process.

Diffusion of Doping Additives

One method of expanding the bandwidth of fiber-optic communications links is to use graded-index (GRIN) optical fibers. Calculations of optimum profiles [8, 9] have indicated that a frequency band of several gigahertz may be achieved in such optical fibers 1 km in length. In order to achieve such a bandwidth the refractive index profile $n(r)$ must be accurate to better than $1 \cdot 10^{-4}$. We have assumed that in order to achieve such accuracy in addition to careful monitoring of the fabrication parameters it is also necessary to account for the redistribution of doping additives during the drawing of optical fibers from the preform [10], although it has been stated [11, 12] that this process is not necessary.

The radial distributions of GeO_2 and B_2O_3 doping additives and the refractive index profile were investigated in preforms and optical fibers fabricated by the MCVD process. The distributions were measured on the "Camebax" X-ray microanalyzer with a spatial resolution of approximately 1 μm and a relative accuracy of 5%. The measurement results for one of the samples are shown in Fig. 1. In the case of the preform (Fig. 1a) a deposited layer structure with a sawtooth germanium distribution in each layer is clear. The GeO_2 concentration varied between 10 and 13 mol% (ignoring anomalous dips). The germanium distribution in the optical fiber drawn from this preform (Fig. 1b) was smoother (the GeO_2 concentrations varied between 9.5 and 11.5 mol%) due to diffusion during the drawing process. This was confirmed with better relative accuracy by near field measurements of

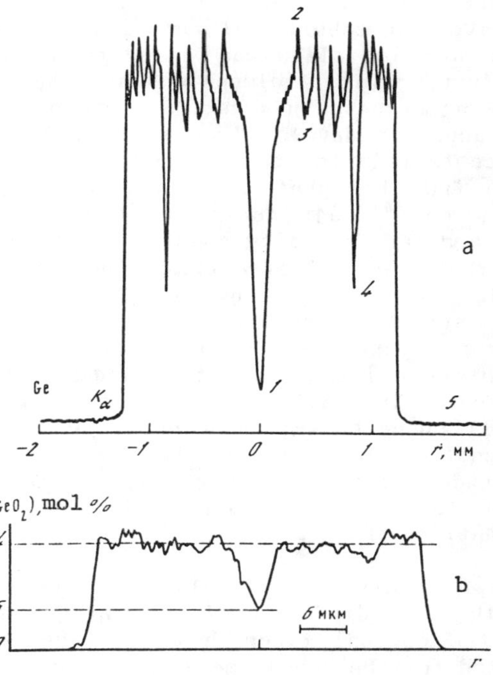

Fig. 1. Radial germanium distribution based on X-ray microanalysis of sample no. 57
a - preform; GeO_2 concentration (mol%): 1 - 1.2; 2 - 13; 3 - 10.5; 4- 5.2; 5 - 0; b - optical fiber

the refractive index profile. It is clear from a comparison of the distributions in Fig. 1 that the refractive index fluctuations are smoothed over lengths of approximately 2 μm in the optical fiber, which corresponds to approximately 100 μm in the preform. The coefficient of diffusion of germanium estimated by the available data is approximately $1 \cdot 10^{-10}$ cm^2/s at $T \approx 2000°C$. Measurement on an optical fiber with a borosilicate reflective cladding revealed that the boron has a diffusion length close to that of germanium and therefore a similar coefficient of diffusion.

In order to determine the influence of germanium diffusion on a refractive index profile of the type

$$n(r) = \begin{cases} n_1 [1 - \Delta (r/a)^{\alpha}], & 0 \leqslant r \leqslant a, \\ n_1 [1 - \Delta] = n_2, & r \geqslant a, \end{cases} \tag{4}$$

where $n(r)$ is the refractive index over a distance r from the optical fiber axis, $\Delta = (n_1 - n_2)/n_1$ is the relative difference of the refractive indices at the center of the core n_1 and in the cladding n_2, a is the core radius; $a \approx 2$ is the profile parameter, and the diffusion equation is solved in cylindrical coordinates. At the points $r = 0$ and $r = a$ this solution allows an analytic estimate. However due to the changes occurring in all dimensions and the diffusion coefficient during the drawing process this estimate may be used only in conjunction with the approach developed in study [13] and by incorporating additional concepts relating to the bulb shape (the region where the fiber is drawn from the preform). Assuming a conical bulb and an equal diffusion coefficient during this entire process of $1 \cdot 10^{-10}$ cm^2/s for standard manufacturing conditions it was determined that at the center of the core the refractive index drops by $4 \cdot 10^{-7}$ while at the core boundary it increases by $4 \cdot 10^{-5}$. In principle in order to compensate the refractive index profile distortion at the boundary it is possible to use convergent boron diffusion from the borosilicate cladding, which reduces the refractive index and has a similar diffusion coefficient. Thus diffusion changes in smooth profile (4) with relatively small concentration gradients are insignificant, although within the deposited layer where the concentration gradient is an order of magnitude greater (see Fig. 1a) and the degree of diffusion equalization of the refractive index fluctuations increases a corresponding amount, reaching $1 \cdot 10^{-3}$.

The derived results are clearly qualitative in nature, although they indicate that the diffusion of the doping impurities occurs primarily during the optical fiber drawing stage and that this effect must be accounted for the development of broadband optical fibers.

OH Hydroxyl Group Diffusion From a Preform Rod

The relatively high diffusion during the drawing process discovered for germanium will also occur with other impurities, including the OH hydroxyl group impurity which causes significant optical losses in optical fibers. As demonstrated in study [14] a concentration of $1 \cdot 10^{-4}$ mass% of OH in an optical fiber core will produce additional losses of 1, 2.4 and 55 dB/km at 0.95, 1.25 and 1.38 μm, respectively. Study [2] has demonstrated that at the 1.55 μm window the OH band wing at 1.38 μm yields additional losses of 0.5 dB/km at an OH concentration of $1 \cdot 10^{-4}$ mass% and of GeO_2 of 19 mol%. Therefore in order to operate in the 1.2–1.6 μm range it is necessary to carefully eliminate the OH from the optical fiber core material.

The contamination cause may be both hydrogen-containing impurities in the raw materials ($SiCl_4$, $GeCl_4$, O_2) as well as diffusion from the vitreous silica preform rod in which the OH concentration may be as high as 0.1 mass% during the MCVD process. Studies [15-17] devoted to this problem indicate, in spite of contradictions that are evidently related to the specific nature of the materials, that OH diffusion from the preform rod is significant. However it is unclear from these studies how significant the diffusion becomes during the drawing process and what thickness of the buffer layer applied to the preform is required for effective protection of the core material. In order to find answers to these questions we investigated the OH concentration distribution in preforms and optical fibers fabricated by the MCVD process [18].

The spectroscopic method was used to experimentally measure the OH distribution in wafers cleaved from the preforms. In view of the complexity of an exact calculation of diffusion as discussed above, a simple exponential law was used to approximate the experimental data

$$C(x) = C(0)\exp(-x/L'), \quad (5)$$

where the distance x is read from the boundary of the deposited material towards the preform axis. The diffusion length L' was 95 ±5 μm.

The optical losses were measured in optical fibers with different buffer layer thicknesses d and the absorption bands of the OH groups were used to determine their effective concentration $C_{эф}$. It turned out that for $d \geqslant 500$ μm the OH concentration stabilized at $0.3 \cdot 10^{-4}$ mass% and this may be due to the presence of hydrogen-containing impurities in the chloride.

On the other hand the value $C_{эф}$ is determined by the radial distributions of OH concentration and the radiation field in the optical fiber. The value of $C_{эф}$ was estimated for exponential distribution (5) with a diffusion length of 95/52 = 1.8 μm (52 is the drawing coefficient of the optical fiber from the preform). According to this estimate in order to reduce $C_{эф}$ to a value of approximately $0.3 \cdot 10^{-4}$ mass% with an OH concentration in the rod of approximately $350 \cdot 10^{-4}$ mass% (as in the experiment) the buffer layer thickness is sufficient for a 170 μm rod. Thus we may conclude that in the optical fiber drawing stage the relative depth of OH diffusion increases by a factor of three compared to the layer deposition and preform shrinkage stage.

Here we should note that since deposition of a thick buffer layer increases the cost of the technology and reduces its productivity level, it may be more advisable to use vitreous silica glass preform tubes with a low OH group concentration or to expose these rods to preliminary thermal treatments in order to reduce the OH concentration in their surface layer.

Germanium Centers in Germanium Silicate Glass

By reducing the transition metal impurity concentration to below 10^{-6} mass% and the hydroxyl groups to 10^{-4}-10^{-5} mass% it is possible to achieve natural optical loss levels in germanium silicate optical fibers in certain regions of the spectrum [1]. At the same time absorption bands produced by germanium are observed in the 0.18-0.4 μm range. Many investigations [19-25] of these absorption bands as well as their resulting luminescence have presented a wide variety of concepts of the nature of these bands. We may reliably assume that they belong to the germanium centers where the germanium ends up in its reduced state with a valency of 2. The purpose of this chapter is to obtain new experimental data on the influence of the technological conditions of the MCVD- and VAD-processes on the concentration and properties of the germanium centers. Since there are significant variations in the temperature and time influences on the preform material in the fabrication of preforms for optical fibers valuable information on the origin of the germanium centers may be obtained from an investigation and comparison of the radial germanium distributions, the UV absorption and luminescence. It is also important to establish the existence or absence of a bond between the germanium centers and the chloride: a technological impurity in glass fabricated by deposition of chlorides [26, 27].

A number of preforms characterized by different fabrication methods, average germanium concentration as well as chloride and OH hydroxyl group concentrations were developed and investigated. Three prototypes in which these differences were most clearly expressed were investigated most extensively. The parameters and features of fabricating such samples are given in Table 2. Sample number 57 was fabricated without a buffer SiO_2 layer and therefore there is a significant differential of the OH concentration at the interface between the core and the cladding [10]. Sample K-114 is anhydrous glass with a typical chloride level for the MCVD process.

Table 2

Parameters of the samples

Sample	Fabrication Technique	Maximum GeO_2 concentration, mol%	Core dia. mm	Buffer layer thickness mm	OH concentration, mass%	Chloride concentration mass%	k_{242}, cm^{-1}	k_{200}, cm^{-1}
No 57	MCVD	13	2.6	-	$5 \cdot 10^{-4}$- $6 \cdot 10^{-2}$	0.1	480	185
K-114	MCVD	5.6	2.1	0.5	$2 \cdot 10^{-4}$	0.1	410	68
C-211	VAD	15.6	4.0	0.4	$2 \cdot 10^{-3}$	0.02	140	42

Sample number S-211 was fabricated by the VAD technique. After vitrification of the porous preform in a vacuum graphite furnace it was expanded to a diameter of approximately 4 mm in an oxygen-propane burner flame and was then "compressed" onto a vitreous silica glass tube. There was no chloride dehydration of the porous preform.

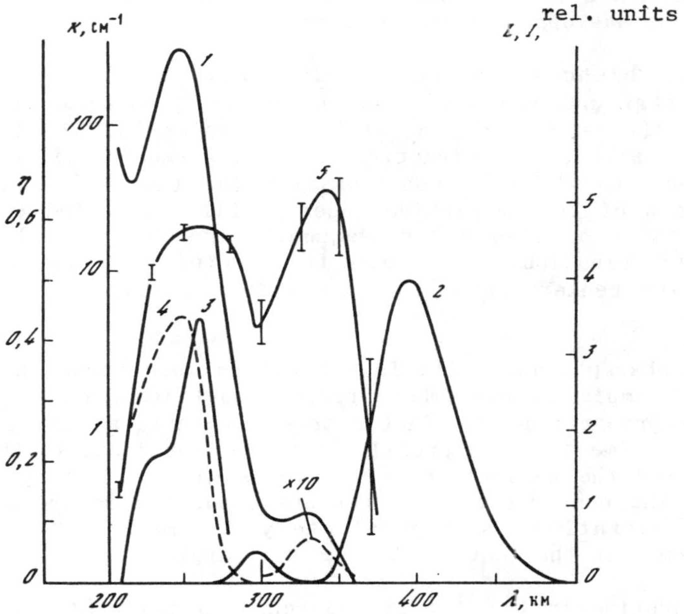

Fig. 2. The absorption spectra $k(\lambda)$ (1), luminescence $L(\lambda)$ (2), luminescence excitation spectra $I(\lambda)$ in the 295 nm (3) and 390 nm (4) bands, and the quantum efficiency $\eta(\lambda)$ (5) in preform No. 57.

The absorption $k(\lambda)$, luminescence $L(\lambda)$, luminescence excitation $I(\lambda)$ and the quantum efficiency $\eta(\lambda)$ spectra were investigated in sample No. 57 and are shown in Fig. 2 together with the luminescence kinetics [23]. Luminescence attenuation in the 390 nm range was exponential with a time constant of 105 μs and the luminescence intensity in the 295 nm band decayed more rapidly than 0.5 μs.

The germanium and chloride distributions were obtained by means of an X-ray microanalyzer with a spatial resolution of approximately 1 μm. We note that this method did not provide information on the chemical bonding of the atoms in the recorded elements. Measurements of the radial luminescence intensity distribution in the 390 nm band were carried out with a spatial resolution from 15-40 μm using various optical schemes. Luminescence was excited by nitrogen laser radiation (λ = 337 nm). Since the absorption coefficient was small ($k_{337} \leqslant 0.3$ cm^{-1}) while the dimensions of the sample d did not exceed 0.4 cm we may assume with a high degree of accuracy that the luminescence intensity is $I_{\text{люм}} \sim 1 - \exp(-k_{337}d) \approx k_{337}d$. The radial distributions of the absorption coefficients were obtained by scanning the samples in a narrow radiation beam exiting the monochromator. Due to the high absorption coefficient at the peak of the 242 nm band in order to improve accuracy the measurements were conducted at different wavelengths within a single band such as 224 nm. The spatial resolution in this case was approximately 100 μm.

Fig. 3 shows the radial distributions of germanium $G(r)$, chloride $c(r)$ and the luminescence intensity reduced to the sample thickness $L(r) = I_{люм}/d \sim k_{337}(r)$, in samples No. 57, K-114, and C-211. We know that the refractive index of germanium silicate glass is proportional to the GeO_2 concentration (or the Ge^{4+} concentration). Measurements of the refractive index profiles revealed that they are in agreement with $G(r)$ and consequently the $G(r)$ profile represents the Ge^{4+} distribution. This would be expected since the concentration of germanium centers is approximately 10^{-3} of the Ge^{4+} concentration [20].

The absorption profile in the 242 nm band shown in Fig. 4a for the S-211 sample is everywhere proportional to L and, consequently, k_{337}; the proportionality factor here is identical for all samples. At the same time the $L(r)$ profile is not proportional to the $G(r)$ profile. Using the extinction coefficients given in study [20] we may determine the concentration of centers absorbing in the 243 nm band. This concentration is approximately $1.2 \cdot 10^{19}$ cm^{-3} in K-114 and $3.5 \cdot 10^{18}$ cm^{-3} in the center of the S-211 sample.

Absorption in the $\lambda < 210$ nm range in study [20, 22] should be considered the natural absorption of GeO_2, i.e,. Ge^{4+}. According to our data in certain samples such as the *S*-211 (Fig. 4b) there is a similarity between $k_{202}(r)$ *and* $G(r)$ (since $\Delta n(r) \sim G(r)$) in the central region of the sample, although they are quite different in the boundary layer with the tube. In other samples (for example, the K-114 sample) $k_{202}(r)$ coincides with $L(r)$. On the basis of these results we may conclude that the absorption bands near 242 and 330 nm and the luminescence bands near 390 nm belong to a single center. The absorption edge in the $\lambda < 210$ nm range represents the superposition of impurity absorption correlating in intensity with the 242 nm band and evidently intrinsic GeO_2 absorption; in this case impurity absorption clearly dominates in the K-114 sample.

The values of L and k_{242} in the S-211 sample and other VAD-samples is, as would be expected from the deposition conditions, less than (by a factor of 3-5) the MCVD samples with similar germanium concentrations through the entire cross-section with the exception of the layer near the interface between the core and the rod where the luminescence intensity is significantly higher. The thickness of the intense luminescence layer is 60-80 μm. Additional investigations of this phenomenon have revealed that the vitrification process in the furnace and the drawing process in the burner flame (hydroxyl groups are introduced into the surface layer in this stage) do not increase L at the surface and the heavily luminescing layer becomes manifest only after the VAD-cladding is "compressed" onto the vitreous silica glass tube. We may conclude from a comparison of these data to the distribution and level of $L(r)$ in the MCVD-preforms (see Fig. 3a, b) that the luminescence intensity increases due to the high-temperature treatments near the surface without direct contact between these surfaces and the burner flame.

According to study [28] the luminescence intensity increases from high-temperature diffusion of hydrogen into the preform resulting from the reduction of GeO_2. We investigated two preforms fabricated by the overlapping rod technique with an identical S-198 VAD-cladding in various melting conditions onto the vitreous silica glass tube. There was no germanium in the surface layer of the 0.2 mm VAD-core. As a result it was discovered that in normal melting where oxygen is injected between the core and the tube (at approximately 1 atm) no increase in *L* was observed, at the same time that melting in a hydro-

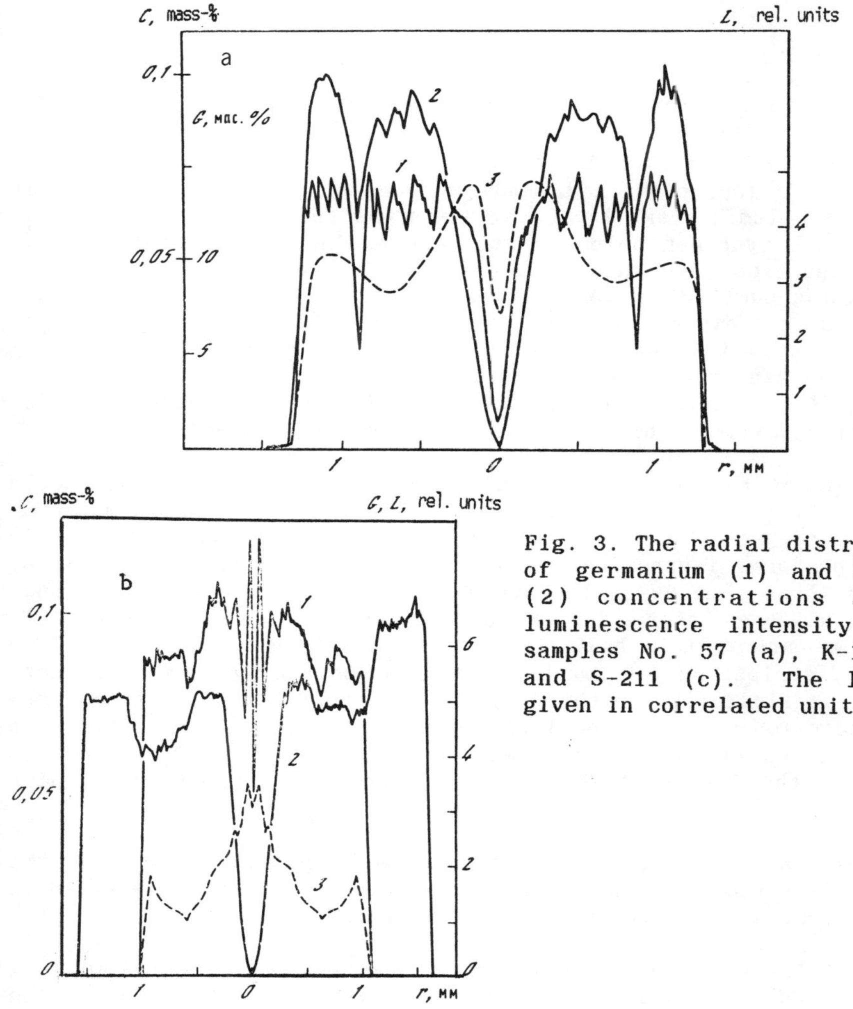

Fig. 3. The radial distributions of germanium (1) and chloride (2) concentrations and the luminescence intensity (3) in samples No. 57 (a), K-114 (b), and S-211 (c). The level is given in correlated units

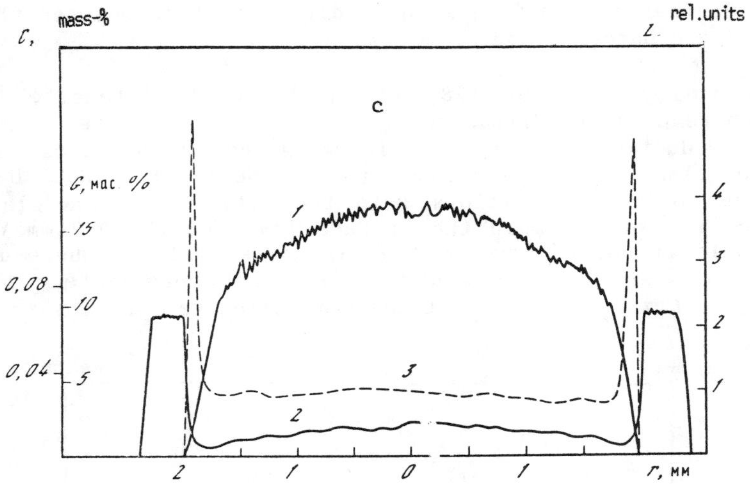

Fig. 3. (continued)

gen atmosphere (approximate 1 atm) produced strong luminescence. In this case the luminescent layer thickness was approximately 70 μm. It is clear that hydrogen helps to increase luminescence. However *L*(*r*) is not proportional to the diffusion distribution of H_2 estimated by the diffusion coefficient of H_2 in vitreous silica glass for our conditions. Such a discrepancy has been observed in the results of study [28]. Evidently the narrow width of the luminescing layer is determined by the significantly lower (with respect to H_2) mobility of the byproducts from the reduction of GeO_2 by hydrogen or by other factors. Additional experiments are needed to clarify this mechanism.

Regarding the chlorine distributions *C*(*r*) (see Fig. 3) we should note that, first, in the VAD-core of the S-211 sample its maximum concentration of approximately 0.02 mass% is significantly less than in the MCVD-preforms (approximately 0.1 mass%). The luminescence intensities in these samples on the average are approximately proportional to *C*(*r*), although there is no correlation between *C*(*r*) and *L*(*r*). Secondly we cannot state that the *c*(*r*) distribution is proportional to the *G*(*r*) distribution, although in samples containing germanium there is a certain similarity in the *C*(*r*) and *G*(*r*) profiles. These data are in contradiction with the conclusions from study [27]. Third, we note the significant breadth of the central valleys in the *C*(*r*) profiles compared to the *G*(*r*) profiles which is evidently related to the high volatility of the chloride and its compounds.

Therefore it has been demonstrated experimentally that the material dispersion does not change from various influences exerted in the technological optical fiber fabrication process.

Quantitative date are derived on the spatial distributions of the doping additives and their influence on the drawing process. The spatial distributions of the hydroxyl groups are investigated and

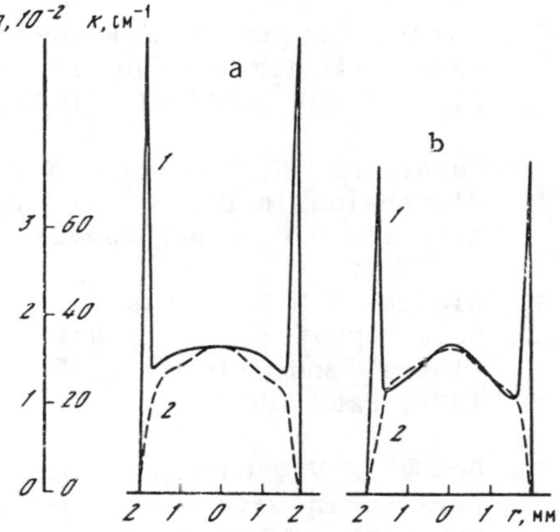

Fig. 4. Radial distributions of the absorption coefficients in the 24 2nm band and in the λ < 210 nm region and the refractive indices in the S-211 VAD-preform
a: 1 - k_{224}; 2 - Δn; b: 1 - k_{202}; 2 - Δn

their depth of diffusion is established while recommendations are developed on buffer layer thickness and as a result the optical losses attributable to the hydroxyl groups are reduced by two orders of magnitude in optical fibers.

The interrelationship between the absorption and luminescence bands was established based on an investigation of the spectral-luminescence characteristics of the germanium centers in germanium silicate glass and on their radial distributions; the formation conditions of the germanium centers were determined with greater accuracy. The technological chlorine impurity which has a comparable concentration is evidently not related to them. The difference in the cross-sectional concentrations of the germanium flaws on the chlorine in the MCVD- and VAD-samples may be qualitatively explained by the specific nature of the processes: the MCVD process occurs at a higher temperature which aids in the reduction of germanium, and with a significantly higher concentration of chlorine in the gas phase, which is responsible for the higher chlorine concentration in the glass.

In conclusion the author wishes to express his gratitude to Ye.M. Dianov, V.B. Neustruev, A.V. Belov, A.V. Nikolaychik, S.V. Lavrishchev, A.N. Gur'yanov, S.M. Mazavin, V.F. Khopin, D.D. Gusovskom, and to many other staff members at the Institute of General Physics of the Academy of Sciences and the Institute of Chemistry of the Academy of Sciences of the USSR for their invaluable assistance in this study.

Bibliography

1. Mija, T., Terunuma, Y., Hosaka, T., et al. "Ultimate low-loss single-mode fiber at 1.5 μm" ELECTRON. LETT., 1979,V. 15, p. 106-109.

2. Belov, A.V., Dianov, E.M., Gurjanov, A.N., et al. "On absorption in GeO_2-doped fused silica fibers" Ibid., 1982, V. 18, p. 836-837.

3. Presby, H.M., Kaminov, I.P. "Binary silica optical fibers: refractive index and profile dispersion measurements" APPL. OPT., 1976, V. 15, p. 3029-3036.

4. Luther-Davies, B., Payne, D.N., Cambling, W.A. "Evaluation of material dispersion in low-loss phosphosilicate-core optical fibers" OPT. COMMUNS., 1975, V. 13, p. 84-88.

5. Kobayashi, S., Shibata, S., Shibata, N., et al. "Refractive index dispersion in doped fused silica" 2nd Intern. Conf. Integrat. Optics and Opt. Fiber Commun. Tokyo, 1977, p. 309-312.

6. Sladen, F.M.E., Payne, D.N., Adams, M.J. "A method for the direct measurement of the inhomogeneous dispersion parameter in optical fibers" 2nd Intern. Conf. Opt. Fiber Transmission, Williamsburg, 1977, Pap. TUE2-1.

7. Belov, A.V., Guryanov, A.N., Dianov, Ye.M., et al. "Material dispersion in vitreous silica glass-based optical fibers" KVANTOVAYA ELEKTRON., 1978, V. 5, p. 695-698.

8. Olshansky, R., Keck, D.B. "Pulse broadening in graded index optical fibers" APPL. OPT., 1976, V. 15, p. 483-491.

9. Okoshi, T. "Optimum profile design of optical fibers and related requirements for profile measurements and control" 2nd Intern. Conf. Integrat. Optics and Opt. Commun., Tokyo, 1977, P. 391-394.

10. Gur'yanov, A.N., Dianov, Ye.M., Lavrishchev, S.V., et al. "Investigation of the material structure of germanium- and boron-doped vitreous silica glass-based preforms and optical fibers" KVANTOVAYA ELEKTRON., 1979, V. 6, p. 2109-2116.

11. Presby, H.M., Standly, R.D., MacChesney, J.B., et al. "Material structure of germanium-doped optical fibers and preforms" BELL SYST. TECH. J., 1975, V. 54, p. 1681-1692.

12. O'Connor, P.B., MacChesney, J.B., Presby, H.M., et al. "Material structure of germanium-doped optical silica, graded index optical fibers" J. AMER. CERAM. SOC. BULL. 1976, V. 55, p. 513-516, 519.

13. Tatarintsev, B.V. "Diffusion at the core/cladding interface in the optical fiber drawing process" FIZIKA I CHIMIYA STEKLA, 1984, V. 10, p. 461-467.

14. Kaiser, P., Tynes, A.R., Astle, H.W., et al. "Spectral losses of unclad vitreous silica and soda-lime silicate fiber" J. OPT. SOC. AMER., 1973, V. 63, p. 1141-1148.

15. Ainslie, B.J., France, P.W., News, G.R. "Water impurity in low-loss silica fiber" MATER. RES. BULL., 1977, V. 12, p. 481-488.

16. Horiguchi, M., Kawachi, M. "Measurement technique of OH-ion distribution profile in rod preform of silica-based optical fiber waveguides" APPL. OPT., 1978, V. 17, p. 2570-2574.

17. Nasau, K. "The diffusion of water in optical fibers" MATER. RES. BULL., 1978, V. 13, p. 67-76.

18. Belov, A.V., Gur'yanov, A.N., Devyatykh, G.G., et al. "Fabrication of optical fibers with a low OH concentration in the core by chemical deposition from the gas phase", FIZIKA I CHIMIYA STEKLA, 1982, V. 8, p. 97-100.

19. Garino, C.V. "Oxyde de germanium ivitrex: Silica vitreuse" VERRES ET REFRACT, 1958, V. 5, p. 313-323.

20. Schultz, P.C. "Ultraviolet absorption of titanium and germanium in fused silica" Proc. XI Intern. Congr. Glass. Prague, 1977, V. 3, p. 155-163.

21. Trukhin, A.N., Boganov, A.G., Praulin'sh, A.M. "The problem of the nature of luminescent centers (396 and 280 nm) in vitreous silica glass", FIZIKA I CHIMIYA STEKLA, 1979, V. 5, p. 346-353.

22. Yuen, M.J. "Ultraviolet absorption studies of germanium silicate glasses" APPL. OPT., 1982, V. 21, p. 136-149.

23. Gur'yanov, A.N., Gusovskiy, D.D., Dianov, Ye.M., et al "Luminescence parameters in germanium-dioxide doped vitreous silica glass" DOKL. AN SSSR, 1982, V. 246, p. 90-93.

24. Amosov, A.V., Petrovskiy, G.T. "'Oxygen vacancy' flaws in vitreous silica glasses" DOKL. AN SSSR, 1983, V. 268, p. 66-68.

25. Skuja, L.N., Trukhin, A.N., Plaudis, A.E. "Luminescence in germanium-doped glassy SiO_2" PHYS. STATUS SOLIDI (a), 1984, V. 84, p. K153-K157.

26. Amosov, A.V., Dorfman, G.A., Zhakhov, V.V.,et. al, "The role of technological hydrogen and chlorine impurities in the radiation processes in vitreous silica glass" FIZIKA I KHIMIYA STEKLA, 1983, V. 9, p. 741-745.

27. Hanafusa, H., Tajuma, Y. "Chlorine concentration profiles in silica fibers" ELECTRON. LETT., 1984, V. 20, p. 178-179.

28. Phylen, D.L., Anderson, W.T. "Analysis of the fluorescence method of profiling single-mode optical fiber preforms" Techn. Digest 5th Topic. Meeting Opt. Fiber Commun., Phoenix, 1982, p. 66-67.

Nonlinear Effects in Monomode Optical Fibers Under Nano- and Picosecond Laser Pumping

Ye.M. Dianov, A.Ya. Karasik, A.V. Luchnikov, P.V. Mamyshev, A.M. Prokhorov

Abstract: Results are given from an experimental investigation of the nonlinear propagation of high-power ultrashort pulses on optical fibers. Pulse propagation is accompanied by self-phase modulation, Raman scattering or Brillouin scattering. Methods of "soliton" pulse compression in the negative group velocity dispersion range are analyzed. A mechanism for forming narrow pulses at the Stokes frequency is proposed that is based on Raman gain from a probe wave self-induced by self-phase modulation. A method is then developed for zero pedestal pulse compression that has made it possible to compress 30-ps pulses from a parametric oscillator by a factor of 150 times. The 200-fs pulses tunable over 1.5-1.67 μm have a power 60 times greater than that of the pump pulses.

The possibility of the influence of acoustic wave diffraction on the generation of nanosecond pulses and the stimulated Brillouin scattering threshold is discussed. Investigations of "traveling" stimulated Brillouin scattering have revealed that the duration of the Stokes pulse exceeds the duration of the pump pulse and may be determined by the length of the medium. A correlation is discovered between the mean duration of the fluctuation Brillouin scattering spikes and the hypersonic relaxation time. The prospects for investigating "traveling" stimulated Brillouin scattering are considered.

Introduction

The consistent and even expanding interest in the formation of ultranarrow light pulses in recent years is associated with their application in both fundamental research and in spectroscopy and in the investigation of high-speed processes in physics, biology and in applications such as broadband fiber-optic communications links [1, 2]. Progress in laser technology has made it possible to develop a number of promising ultranarrow light pulse generators. These include picosecond solid-state and passive and active mode locked gas lasers that have in turn led to the development of synchronously-pumped dye and crystal lasers and parametric light generators as well as subpicosecond colliding pulse lasers [1-4].

Another alternative approach to the formation of ultranarrow light pulses involves nonlinear transformations of laser radiation. These include such familiar nonlinear effects as stimulated Raman scattering, stimulated Brillouin scattering, self-phase modulation of radiation, etc. The recent development of low-loss fused silica glass optical fibers (with losses of a few decibels per kilometer or lower) has provided a new stimulus to such research. This is due primarily to the many unique properties of optical fibers such as the virtually constant transverse modal field configuration in the optical fiber over an extensive (compared to bulk media) interaction length of the radiation with the medium; the sharp drop in the radiation power threshold for nonlinear effects; the extended spectrum of mode resonances in fused silica; the broad spectral transparency range of fused vitreous silica glass from 0.4 to 2 μm; and the existence of bands in this range with positive and negative dispersion of the group velocities of light ($\partial v_{\Gamma p}/\partial \lambda$) and a number of other properties. These factors have made it possible not only to achieve a number of new nonlinear effects in optical fibers, but also to investigate new features of familiar effects [5-7].

1. Self-Steepening of Light Pulses

One of the most promising nonlinear optical techniques of generating ultranarrow light pulses is compression of initial laser pulses based on the self-phase modulation of radiation in a nonlinear medium [8]. This method makes it possible not only to generate ultranarrow light pulses but also to concentrate a larger portion of the pump pulse energy in a compressed pulse.

We will examine this method in greater detail. The self-phase modulation effect is related to the nonlinearity of the refractive index of the medium which depends on the radiation intensity:

$$n(|E|^2) = n_0 + n_2 |E^2|/2, \tag{1}$$

where n_0 is the refractive index in the linear approximation; E is the electric field strength of the light wave; n_2 is the nonlinear refractive index.

Formula (1) is written assuming infinitely fast relaxation of the nonlinearity of the medium. With propagation of the light pulse $|E(t)|^2$ (Fig. 1a) through such a medium the various segments of the pulse will experience an additional phase shift:

$$\delta\Phi(t) = \frac{2\pi}{\lambda} L \frac{n_2}{2} |E(t)|^2, \tag{2}$$

where L is the length of the nonlinear medium; λ is the radiation wavelength in a vacuum. Expression (2) is written assuming low dispersion of the medium, i.e., the pulse waveform is assumed to remain constant during propagation through the medium.

The frequency shift $\delta\omega(t)$ corresponding to the additional phase shift $\delta\Phi(t)$ in the various pulse sections is represented by the time derivative of the phase (Fig. 1b)

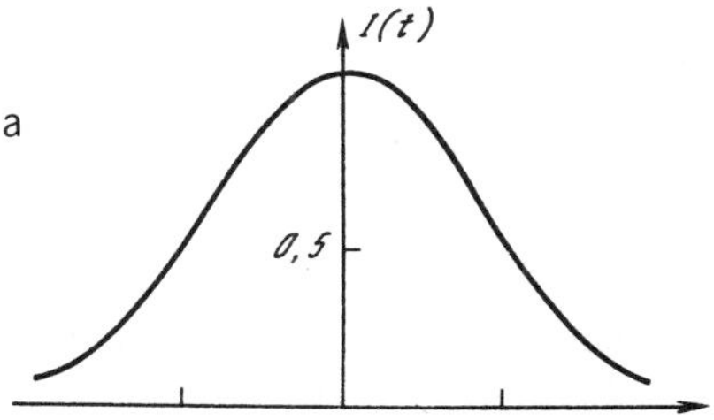

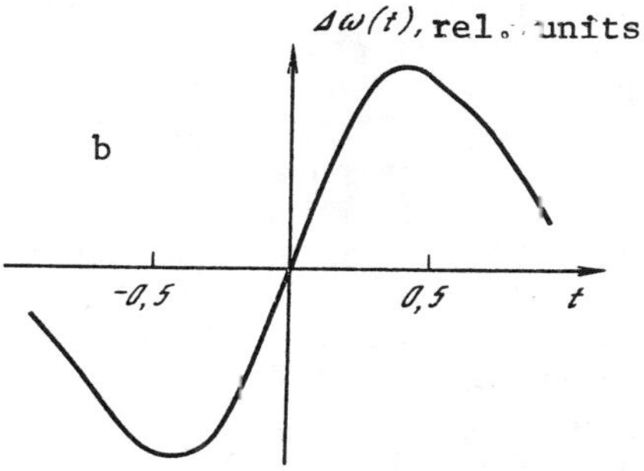

Fig. 1. Pulse envelope $I(t)$ and frequency modulation of the pulse $\Delta\omega(t)$ from pulse propagation through a nonlinear medium from the self-phase modulation effect

$$\delta\omega(t) = -\frac{d\delta\Phi(t)}{dt} = -\frac{2\pi L}{\lambda}\frac{n_2}{2}\frac{d|E(t)|^2}{dt}, \tag{3}$$

while the frequency spectrum of the pulse after its self-phase modulation is given by the Fourier transform

$$F(\omega) = \frac{1}{2\pi}\int P(t)^{1/2}\exp[i\delta\Phi(t)]\exp[-i(\omega-\omega_0)t]\,dt, \tag{4}$$

where $P(t)$ is pulse power; ω_0 is the central frequency of the laser radiation.

The spectral broadening $\Delta\omega$ of the pulse from self-phase modulation is greater the higher the maximum phase shift $\Delta\Phi_{\text{макс}}$:

$$\Delta\Phi_{\text{макс}} = \frac{2\pi L}{\lambda}\, n_2 \frac{|E(0)|^2}{2}. \tag{5}$$

Thus, for a Gaussian pulse the spectral broadening is equal to

$$\Delta\omega = 0{,}86\,\Delta\omega_0\Delta\Phi_{\text{макс}}, \tag{6}$$

where $\Delta\omega_0$ is the initial spectral pulse width.

Therefore the spectrum of the initial pulse expands due to the self-phase modulation effect and the pulse acquires frequency modulation and, as we see from (3) and Fig. 1b when $n_2 > 0$ the frequency shifts to the Stokes region at the trailing edge of the pulse ($\delta\omega(t) < 0$), and shifts to the anti-Stokes region ($\delta\omega(t) > 0$) at the trailing edge of the pulse. If such an FM pulse is transmitted through a delay line with negative dispersion (the optical length increases with increasing wavelength), the "blue" frequency components catch up with the "red" frequency components and thus the initial pulse may be compressed to a degree determined by the inverse spectral width of the pulse after its self-phase modulation. Study [9] proposed that a pair of diffraction gratings be used as such a dispersion delay line for compression of frequency-modulated laser pulses.

The light pulse compression technique described above employing self-phase modulation in a nonlinear medium was previously proposed in study [10] and was used in several experiments that employed fluids as the nonlinear medium [11-12]. In order to achieve significant spectral broadening by the self-phase modulation technique these experiments required the use of radiation power levels that are accompanied by self-focusing of emission [13-15]. In practice it is impossible to achieve significant spectral broadening of radiation in a nonwaveguide medium by self-phase modulation without self-focusing of emission, since the length of nonlinear interaction between the radiation and the medium, determined by the diffraction divergence, is rather short and the more sharply the radiation is focused in the medium, the shorter the interaction length.

Indeed, assume the radiation is focused on a spot of radius r (r is the beam radius where the intensity decays by a factor of e). Then the length of nonlinear interaction of the radiation with the medium l is (l is equal to twice the length over which the cross-sectional area of the beam expands by a factor of 2 due to linear diffraction)

$$l = n_0 4\pi r^2/\lambda. \qquad (7)$$

Here λ is the radiation wavelength in a vacuum; n_0 is the refractive index. Subject to the expression for the critical self-focusing power [15]

$$P_{\text{сф}} = c\lambda^2/(32\pi^2 n_2) \qquad (8)$$

(c is the speed of light) and expressions (5), (6) we obtain an estimate of the maximum possible spectral pulse broadening due to self-phase modulation without self-focusing in the nonwaveguiding case

$$\frac{\Delta\omega}{\Delta\omega_0} \approx \Delta\Phi_{\text{макс}} < \frac{1}{2}\,\frac{P_{\text{сф}}}{\pi r^2}\,\frac{8\pi}{n_0 c}\,n_2\,\frac{2\pi}{\lambda}\,l = 1. \qquad (9)$$

Expression (9) indicates that in a nonwaveguiding medium with diffraction divergence of the radiation without self-focusing the spectral broadening of radiation due to self-phase modulation will not exceed the initial spectral width. For this reason efficient self-phase modulation of radiation without self-focusing was first observed experimentally in optical fibers [16]. Self-focusing complicates subsequent pulse compression: as a rule the radiation beam decomposes into self-focusing filaments, becomes highly nonunimodal and spectral broadening is highly inhomogeneous in the beam cross-section. For these reasons there have not been extensive experiments devoted to picosecond light pulse compression using self-phase modulation in bulk nonlinear medium in self-focusing conditions [11, 12].

A significantly different situation occurs when monomode silica glass optical fibers are used as the nonlinear medium where the spatial intensity profile (to powers of the order of $P_{C\Phi} \approx 10^6$ w, where $P_{C\Phi}$ is the critical self-focusing power in vitreous silica glass) is constant along the entire optical fiber length and is virtually unaffected by nonlinearities [17, 18]. In this case the radiation is localized in 3–10 μm cross-sections over optical fiber lengths that may in principle be as high as tens of kilometers. Thus, the development of high-quality monomode optical fibers that made it possible to achieve near-linear FM and homogeneous spectral broadening in the beam cross-section by means of the self-phase modulation effect without self-focusing prompted a renaissance of the ultranarrow light pulse compression method described above. Varying the optical fiber length over a significant range makes it possible to achieve efficient self-phase modulation of radiation over a broad dynamic range of pump power (from a few watts to tens of kilowatts) which significantly expands the application of relatively low-power, broadly-tunable pulsed lasers that may also operate at a high repetition rate. Today this method is used to achieve efficient compression of radiation pulses from synchronously-pumped dye lasers [19], colliding pulse dye lasers [20, 21] and for compression of first and second harmonic pulses from CW-pumped active mode-locked YAG:Nd^{3+} lasers operating in a double-Q-switched regime [22] and a CW state [23, 24].

Therefore as the waves propagate in a cubic nonlinear medium self-action effects arise due to the dependence of the wave velocity on its amplitude. The most familiar example is self-focusing of light pulses. Another form of self-action is the space-time deformation of the amplitude and frequency envelopes of a light pulse caused by the combined action of self-phase modulation and dispersion of the medium [25–28]. Since silica glass optical fibers have negative dispersion of the group velocities ($dv_{\Gamma p}/d\lambda < 0$) in the spectral range $\lambda > 1.3$ μm light pulses traveling through an optical fiber may undergo self-steepening due to the combined influence of self-phase modulation and dispersion $dv_{\Gamma p}/d\lambda < 0$ [25, 29] and therefore the optical fiber itself without an external dispersion delay line will function as a light pulse compressor.

The nonlinear Schrodinger equation

Normally the theoretical analysis of nonlinear propagation of light pulses in optical fibers is carried out with the following assumptions [30, 31]:

1) the electric field is taken as a transverse field and is sought as

$$\psi(x, y, z, t) = \tfrac{1}{2}\, \chi(x, y)\{\epsilon(z, t)\exp[i(\omega t - kz)] \quad \text{complex conjugation}\} \tag{10}$$

where $\chi(x, y)$ describes the transverse spatial field distribution in the linear approximation ($\chi(0, 0) = 1$), while a slowly varying amplitude of the temporal envelope $\varepsilon(z, t)$ is subjected to a nonlinearity (z is the coordinate on the optical fiber axis, x, y are the transverse coordinates, t is time);

2) the examination is carried out in the second approximation of dispersion theory, i.e., the expansion of the wave number $k(\omega)$ in powers of $(\omega - \omega_0)$ accounting for terms of second order of smallness.

In this case the effective electric field averaged over the transverse coordinates $E(z, t)$ satisfies the nonlinear Schrodinger equation

$$i\frac{\partial E}{\partial z} = -\frac{1}{2}\frac{\partial^2 k}{\partial \omega^2}\frac{\partial^2 E}{\partial \tau^2} + \frac{1}{2}\frac{n_2}{n_0}k_0|E|^2E - i\gamma E. \tag{11}$$

Here $\tau = (t - z/v_{\text{гр}})$ is time traveling ["co-moving"] with the pulse; $v_{\text{гр}} = \partial\omega/\partial k$ is the radiation group velocity; γ is the coefficient of linear absorption; $k_0 = (2\pi/\lambda)n_0$. The effective electric field is determined as

$$E(z, t) \equiv \epsilon(z, t)\sqrt{<\chi^4>/<\chi^2>}, \tag{12}$$

where the < > infer integration over the transverse coordinates.

The quantity $E(z, t)$ is related to the luminous radiation power $P(z, t)$ in the following manner:

$$|E(z, t)|^2 = \frac{8\pi}{n_0 c}\frac{P(z, t)}{S_{\text{эф}}}, \tag{13}$$

where the effective cross-sectional area of the mode $S_{\text{эф}}$ is equal to [61]

$$S_{\text{эф}} = <\chi^2>/<\chi^4>. \tag{14}$$

Thus the dynamics of light pulse propagation in a optical fiber will be determined by the combined influence of dispersion (the first term in the right half of equation (11)) and self-phase modulation (the second term in the right half of (11)). Equation (11) is written for the case of an optical fiber conserving linear polarization; if depolarization does not occur then n_2 in equation (11) must be substituted by the polarization-averaged value of 5/6 n_2 [18, 32].

Equation (11) assumes an infinitely fast relaxation of nonlinearity n_2. In the case of a fused silica-based optical fiber such an assumption is quite justified, since in glasses such a nonlinearity is determined primarily by the electron contribution [33]. We note that the finite relaxation time of the nonlinearity (for example, in the case of nonlinear fluids) may give rise to shock wavefronts of the pulse envelopes [27, 34]. In the case of femtosecond pulse durations the assumption of a slow variation of the temporal pulse envelope becomes invalid, since it is necessary to account for additional terms responsible for the asymmetrization of the temporal envelope of the channeled pulse, specifically the formation of envelope shock waves in equation (11) and similar equations [26, 35-39].

A further examination of equation (11) depends on the sign of $\partial^2 k/\partial\omega 2$. The $\partial^2 k/\partial\omega^2 < 0$ corresponds to negative group velocity dispersion ($\partial\upsilon_{\Gamma p}/\partial\lambda < 0$), while the case $\partial^2 k/\partial\omega^2 > 0$ corresponds to positive group velocity dispersion ($\partial\upsilon_{\Gamma p}/\partial\lambda > 0$). For fused silica glass-based optical fibers the first case is realized at wavelengths $\lambda > 1.3$ μm while the second case is realized at $\lambda < 1.3$ μm [40]. We note that optical fiber dispersion is not identical to the material dispersion of the optical fiber, since it is determined by the sum of the material and waveguide dispersions [41]. For this reason the wavelength with zero optical fiber dispersion ($\partial^2 k/\partial\omega^2 = 0$) may in principle be varied over a broad range by, for example, varying its refractive index profile. We note that when $\partial^2 k/\partial\omega^2 = 0$ equation (11) may contain higher order dispersion terms ($\partial^3 k/\partial\omega^3$, etc.) [42, 43].

Pulse propagation in a optical fiber in the spectral range of negative group velocity dispersion

In the negative group velocity dispersion range (i.e., for $\partial^2 k/\partial\omega^2 < 0$) when $\gamma = 0$ equation (11) has soliton solutions [42-45] for initial pulses of the type

$$E(z=0,\tau) = \frac{1}{\Delta t_0}\sqrt{\left|\frac{\partial^2 k}{\partial\omega^2}\right|\frac{\lambda}{\pi n_2}}\, N \mathrm{sech}\left(\frac{\tau}{\Delta t_0}\right), \tag{15}$$

where N is an integer; $\Delta t_0 = 0.568\tau_0$; τ_0 is the pulse duration at half-amplitude.

The case $N = 1$ corresponds to a fundamental soliton. In this case self-phase modulation producing a frequency shift of the leading edge of the pulse to the "red" frequency range and the trailing edge to the "blue" frequency range (see Fig. 1) and dispersion producing a delay of the "red" frequency components with respect to the "blue" components completely compensate. In this case the pulse travels through the optical fiber without change of waveform. Moreover, the fundamental soliton is stable with respect to small perturbations [30, 44].

For the case of a launched pulse with initial amplitude $E(0, \tau)$ less than the corresponding value for the fundamental soliton (i.e., $N \ll 1$ in (15)), dispersion predominates over nonlinearity and the pulse broadens as it propagates.

When $N > 0$ the nonlinearity predominates in the early pulse propagation stage and the pulse will undergo self-steepening in this stage. Overall the propagation dynamics of such pulses are rather complex, while for soliton solutions (N is an integer and such solutions are called N-soliton pulses or bound states of N solitons [44, 45]) and the initial pulse waveform will be recovered in the soliton period [46]

$$z_0 = \frac{\pi}{2} \frac{\Delta t_0^2}{|\partial^2 k/\partial\omega^2|}. \tag{16}$$

Fig. 2 shows the dynamics of soliton pulse propagation for the case where $N = 3$ in a period z_0 [46].

At a launched pulse power exceeding the fundamental soliton power the pulse will undergo self-steepening in its initial propagation stage (Fig. 2). This effect may be used to generate ultranarrow pulses. Study [47] has carried out calculations of the waveguide length corresponding to maximum pulse compression and the degree of compression and waveform of a compressed pulse for launched pulses of the type in (15). Study [47] finds that the maximum self-steepening length drops with growth of N, and the compression ratio $\tau_0/\tau_{сж}$ increases and at high N is approximated by the value

$$\tau_0/\tau_{сж} \approx 4.1\ N. \tag{17}$$

However with this method a low-intensity, broad pedestal is formed under the compressed, high-intensity peak; the pedestal contains significant energy and the portion of pedestal energy grows with an increase in the compression ratio. Therefore as indicated by the calculations when $N = 20$ the compression ratio is $\tau_0/\tau_{сж} \approx 82$, although only approximately 17% of the initial pulse energy is contained in the compressed pulse spike, while the remaining 83% of the energy "escapes" to the broad pedestal. The existence of such a pedestal not only degrades the energy characteristics of the compressed pulses, but also complicates their practical use. In spite of the extensive theoretical studies devoted to the nonlinear dynamics of ultranarrow pulse propagation in a optical fiber in the spectral range of negative group velocity dispersion revealing extensive interest in this problem, until recently only a single group of researchers has carried out the corresponding experimental investigations [46, 48-50]. This is due to the problem of building stable picosecond pulse sources without a substructure in the $\lambda > 1.3$ μm spectral range and of fabricating low-loss monomode optical fibers. The authors of studies [46, 48-50] employed in their experiments an F^{2+} color center picosecond laser employing an NaCl crystal at $\lambda = 1.55$ μm operating at $T = 77$ K synchronously-pumped by the first harmonic from an active mode-locked CW YAG:Nd^{3+} laser. Study [50] obtained a 27-fold compression of 7 ps pulses from this laser in a monomode optical fiber section 100 m in length. As predicted by theory the 260 femtosecond compressed pulses had a broad pedestal.

Fig. 2. Computer calculation within the framework of equation (11) of the $|E(z, \tau)|^2$ envelope dynamics of a three-soliton pulse (the bound state of three solitons) in period z_0
The pulse undergoes self-steepening in the initial propagation stage; the optical fiber is of the length where pulse compression is maximized. The pulse recovers its initial waveform in soliton period z_0.

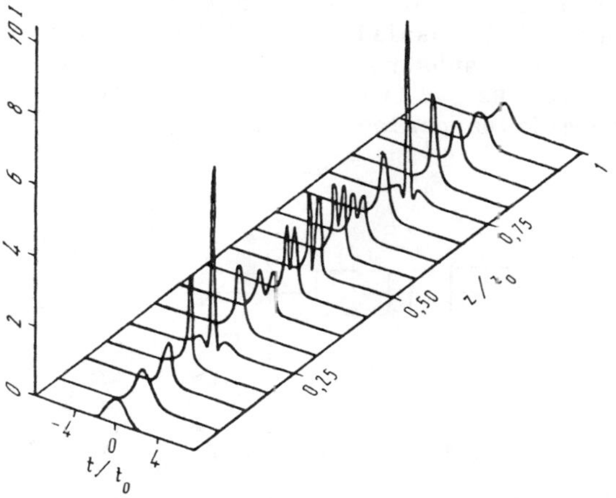

The purpose of the present study was to investigate the possibility of determining the high compression ratios of ultranarrow light pulses in monomode optical fibers in the negative group velocity dispersion range and of obtaining compressed pulses without a pedestal with a high energy conversion factor.

Experimental technique and instrumentation

The following requirements are imposed on the radiation sources used to investigate nonlinear ultranarrow light pulse propagation through optical fibers: 1) the generated picosecond pulses cannot have a temporal substructure; 2) the pulse power must be sufficient for efficient nonlinear processes; 3) the pulses must have a high pulse repetition rate and high stability for reliable recording of relatively low-energy pulses; 4) the radiation sources must have an extensive tuning range for investigations in the spectral ranges of interest to us.

Today the most widely employed radiation sources are continuously-pumped active mode-locked YAG:Nd^{3+} lasers operating both in CW and in active double Q-switched conditions. In the latter case the laser radiation consists of trains of high-power ultranarrow pulses ~100 ps in duration with a high repetition rate (up to 50 kHz). Specifically the first and second harmonic radiation from both the CW and the double Q-switched lasers may be used successfully for synchronous pumping of picosecond color center and dye lasers [4, 49-53] which represent serious competition to traditional gas lasers used for such applications.

The present study employs a continuously tunable (λ = 0.74-1.9 μm) parametric oscillator based on a nonlinear barium-sodium niobate crystal ($Ba_2NaNb_5O_{15}$) synchronously pumped by the second harmonic of a continuously pumped active mode-locked and Q-switched YAG:Nd^{3+} laser as the radiation source to investigate the nonlinear propagation of ultranarrow pulses on optical fibers [22, 53]. The parametric oscillator emitted trains of ultranarrow pulses without a temporal substructure 30 ps in duration with a train repetition rate of 100 Hz, a train half-amplitude duration of 100-250 ns and a pulse repetition period of 4 ns.

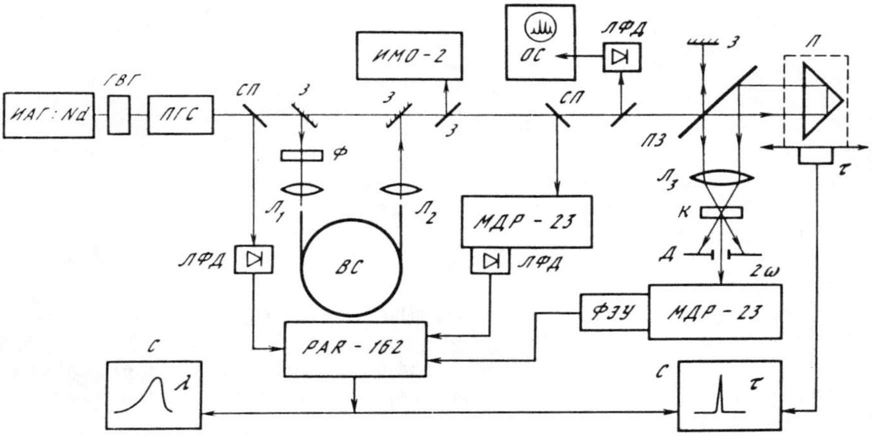

Fig. 3. Experimental set-up for investigating nonlinear ultranarrow pulse propagation conditions in an optical fiber in the spectral range corresponding to negative group velocity dispersion
YBT - $LiIO_3$ crystal for second harmonic generation of $YAG:Nd^{3+}$ laser radiation; $Л_1$-$Л_3$ - lenses; ВС - optical fiber; СП - beam splitting plates; З - mirror; F - light filters; ПЗ - semitranasparent mirror; П - micrometer rectangular prism with displacement transducer; K- second harmonic crystal 1 mm in thickness; Д - diaphragm; ЛфД- avalanche photodiodes; ИМО-2 - mean laser power meter; ОС - S1-75 oscillograph; МДР-23 - grating monochromators; PAR-162 - gated integrator; C - recorders; ИАУ:Nd - $YAG:Nd^{3+}$

The experimental set-up for investigating the nonlinear propagation of ultranarrow light pulses through an optical fiber in the negative group velocity dispersion range of the optical fiber is shown in Fig. 3. Radiation from the parametric oscillator is launched into the optical fiber by means of lens $Л_1$ with a focal distance of 1 cm. A monomode optical fiber of L = 250 m was selected for the experiments (core diameter $2a \approx 10$ μm with a difference of the refractive indices of the core and cladding of $\Delta n \approx 4 \cdot 10^{-3}$).

The radiation was taken from the optical fiber exit and impinged through collimating lens $Л_2$ either the monochromator used to measure the radiation spectrum or the nonlinear lithium iodate crystal ($LiIO_3$) correlator used to measure the temporal characteristics of the laser radiation. The zero-background intensity autocorrelation functions were measured with noncollinear second harmonic generation by crystal K 1 mm in thickness. The monochromator at the correlator output made it possible to measure the temporal characteristics of the various frequency components of the emission exiting the optical fiber [54, 55]. The frequency components are selected by monochromator tuning and by rotating crystal *K* to the appropriate sync angle.

The radiation spectra and correlation functions were recorded on recorders *C* through a gated integrator (strobe of ~10 ns). This made it possible to carry out measurements on pulses selected from a specific portion of the parametric oscillator pulse train.

The power characteristics of the laser radiation were based on the measured mean radiation power level by means of a power meter (IMO-2) and the light pulse duration measured by the correlator. Neutral light filters were used to measure the radiation fed to the optical fiber. The power level of the injected radiation could also be varied by altering the feed conditions to the optical fiber (the angle of radiation injection to the optical fiber, tuning of the entrance end of the optical fiber in the focal region of lens $Л_1$).

This experimental assembly made it possible to measure the correlation functions of the various frequency components of the laser radiation together with the radiation spectra at the entrance and exit of the optical fiber for different injected laser radiation power levels.

Experimental results

The propagation of ultranarrow pulses through an optical fiber is accompanied by self-phase modulation producing spectral pulse broadening (see Fig. 1). Fig. 4 shows the spectra of the exiting radiation with a successive increase in the parametric oscillator emission power level injected to the optical fiber at $\lambda = 1.54\ \mu m$. When the spectral broadening $\Delta\lambda$ of pulses of duration τ_0 resulting from self-phase modulation in an optical fiber with dispersion D and length L is small, the condition $D\Delta\lambda\ L/2 \ll \tau_0$ is satisfied and we may assume that the waveform and duration of pulses at the optical fiber exit are virtually identical to those of the initial pulses. In this case we may use expression (2) for the self-phase modulation in a zero-dispersion medium for the spectral broadening value. The injected pulse power for the case of Fig. 4a was calculated by the measured radiation spectral width at the optical fiber exit ($\Delta\nu \approx 8\ cm^{-1}$) at $P = 50$ w. In the case of Fig. 4c the power $P = 80$ w was determined without one of the calibrated neutral light filters Φ with a known transmission factor; these filters are normally located in front of the injection lens $Л_1$.

Determinations of the power level of pulses launched into the optical fiber based on the mean power level registered on the ИМО-2 instrument was limited by the sensitivity of the ИМО-2 instrument and was possible for the case $P > 200$ w. Thus a power of $P = 900$ w was measured for the case in Fig. 4d directly by means of ИМО-2.

As we see from Fig. 4 with successive increases in the power level of the $\tau_0 \approx 30$ ps light pulses launched into the optical fiber the spectrum of radiation exiting the optical fiber expands, and at $P = 80$ w (Fig. 4c) a longwave Stokes wing appears in the previously symmetrical spectrum (Fig. 4a, b). The relative portion of energy contained in the wing increases with further increase in the power injected to the optical fiber, and when $P = 900$ w it contains up to 50% of all energy contained in the radiation exiting the optical fiber.

The autocorrelation function at the pump frequency (Fig. 5a) and the autocorrelation function corresponding to the Stokes spectral wing for the frequency shift with respect to the pump frequency $\Delta\nu \geqslant 55$ cm^{-1} (Fig. 5b) measured at $P = 900$ w are shown in Fig. 5. The autocorrelation function (Fig. 5b) corresponding to the Stokes spectral wing was recorded both through the monochromator (the spectral width of the slit was approximately 5 nm) and without the monochromator. In the latter case the "spectral width of the slit" was the width of the phase synchronism of second harmonic crystal K (see Fig. 3). (The

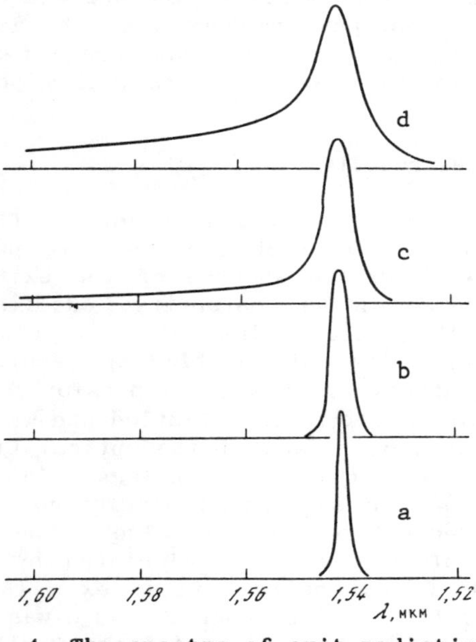

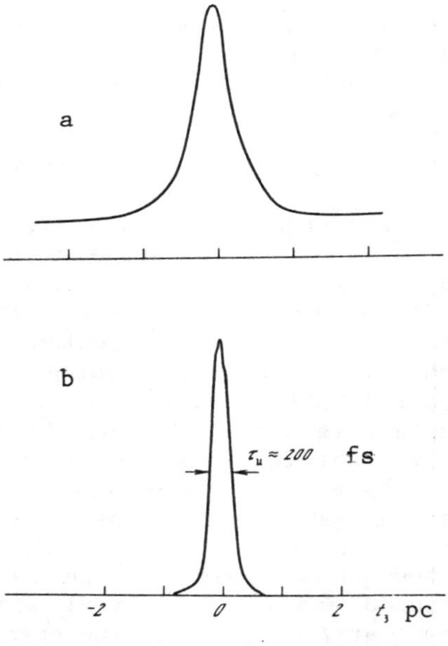

Fig. 4. The spectra of exit radiation from the optical fiber with different injected radiation power levels

Fig. 5. Intensity autocorrelation functions measured at $P \approx 900$ w at the optical fiber exit
a - at the pump frequency (the wings of the function extend to $t_3 = \pm 50$ ps); b - at the Stokes frequency with respect to the pump frequency (there is no pedestal in recording the function to $t_3 = \pm 60$ ps)

calculated value of the phase synchronism width for collinear second harmonic generation at $\lambda = 1.6$ μm for the $LiIO_3$ crystal 1 mm in length employed in the autocorrelator was ~33 nm.) The spectral width of the second harmonic of the Stokes spectral wing (see Fig. 4d) in the collinear case measured at half-amplitude is 280 cm^{-1}. In the case where the radiation components with a frequency shift with respect to the center of the pump line less than 50 cm^{-1} did not fall within the "spectral width of the slit", the autocorrelation function shown in Fig. 5b remained virtually the same both when recorded through the monochromator and without the monochromator. The autocorrelation function in Fig. 5b represents a zero-pedestal pulse of duration $\tau \approx$ 200 fs (assuming a pulse waveform of $sech^2$) at the Stokes frequency with respect to the pump frequency. (Measurement accuracy of the autocorrelation function width is 20%.) The energy of this pulse calculated by the spectral waveform (see Fig. 4d) and the launched pulse energy is approximately 10 nJ. Such an energy corresponds to a Stokes pulse power of 55 kw for a duration τ = 200 fs.

Thus, when radiation at λ = 1.54 μm with τ_0 = 30 ps and $P \approx$ 900 w is injected to an optical fiber 250 m in length single pulses τ = 200 fs in duration with an energy of ~11 nJ appear at the Stokes frequency (with respect to the pump frequency). Analogous results have been obtained from pump emission tuning in the 1.5-1.65 μm range.

Discussion of Result

We will now consider the causes behind the Stokes wing appearing in the spectrum of exit radiation (see Fig. 4c, d) and the formation of a pedestal-free powerful narrow pulse in the Stokes region (see Fig. 5b).

Significant spectral broadening of powerful laser pulses has been observed previously in a variety of media, particularly with self-focusing. Various self-phase modulation mechanisms have been proposed to explain these effects together with spectral broadening mechanisms from stimulated parametric four photon processes (see survey 1, as well as studies [54, 56-66]). Self-phase modulation of symmetric pulses with zero self-focusing and preferential spectral broadening to the Stokes region is possible at nonlinearity relaxation times comparable to the pulse duration [8] which does not correspond to our case, since nonlinearity relaxation will be virtually instantaneous in glass. For this reason the shock wave formation mechanism of the light pulse envelopes examined by Ostrovskiy [27] cannot exist in glass.

We propose the following stimulated Raman gain mechanism from the probe wave self-induced by self-phase modulation to explain the appearance of a high-energy Stokes wing in the radiation spectrum in our experiments. As the pump pulse propagates through the optical fiber its spectrum is broadened due to self-phase modulation, and the symmetric spectral form (see Fig. 4a, b) indicates a symmetric pulse waveform (see (3)). The pulse traveling through the optical fiber begins to compress due to the combined influence of self-phase modulation resulting in a frequency shift to the Stokes region at the leading edge of the pulse and to the anti-Stokes region at the trailing edge of the pulse and the negative group velocity dispersion of the optical fiber, thereby forming a narrow, high-intensity spike on a broad pedestal background at the pump frequency (see Fig. 2) as predicted by theory [48]. Estimates based on formulae (15), (16) revealed that in our case for τ_0 = 30 ps the soliton period is $z_0 \approx$ 18 km and the fundamental soliton power $P_1 \approx$ 80 mw. For P = 80 w (see Fig. 4c) the appearance of the Stokes wing in the spectrum is of the order of the soliton $N = \sqrt{P/P_1} \approx 30$, and the optical fiber length corresponding to maximum pulse self-steepening with N = 30 is 200-300 m (the theoretical relation from study [50] was extrapolated). In this case the duration of the compressed pulse against the pedestal background is 120 times narrower than the pulse at the optical fiber entrance (formula (17)).

A remarkable property of fused silica glass is the continuous spectrum of the inhomogeneously-broadened vibrational resonances extending to the Stokes region virtually from 0 to over 1000 cm^{-1}, and even for a Stokes shift of $\Delta\nu$ = 55 cm^{-1} the spectral intensity of spontaneous Raman scattering is only 1.5 times smaller than the maximum intensity for $\Delta\nu$ = 440 cm^{-1} (Fig. 6) [67].

A portion of the spectrum of the compressed pump peak falls within the Raman gain region [68]. For a Gaussian pulse of duration τ = 30 ps/120 = 0.25 ps the spectral density for a frequency shift of $\Delta\nu$ = 55 cm^{-1} is ~0.1 of the corresponding value at the center of the line.

Thus the pump pulse itself generates a rather intense primer at the Stokes frequency and the Raman gain process from this primer rapidly enters saturation. Analysis reveals that the Raman threshold from spontaneous noise is not achieved at our pump powers of P < 1 kw. We note that the primer will not necessarily be formed at the point of maximum pump pulse compression. Fig. 7 gives a qualitative representation of the formation of the Stokes primer before the maximum pulse compression point in the optical fiber. The hatched regions represent the portions of the spectrum and pulse falling within the Raman-gain region.

As the Stokes pulse is formed and propagates through the optical fiber it begins to slow down with respect to the pump pulse due to optical fiber dispersion with $\partial\upsilon_{rp}/\partial\lambda < 0$ (for $\Delta\upsilon$ = 55 cm^{-1} the delay is ~0.3 ps/m) and therefore "eats up" the trailing edge of the pump pulse pedestal. The higher the pump power. the earlier the Raman scattering appears and the more energy can be accumulated by the Stokes pulse.

We should note that this situation where the scattering pulse is continually amplified in the unexhausted pump field is, in principle, comparable to the backward stimulated Raman state which was first investigated theoretically and experimentally in studies [69, 70]. We also note that stimulated Raman scattering with frequency shifts $\Delta\upsilon$ less than 440 cm^{-1} have also been observed experimentally in vitreous silica monomode optical fibers [71].

Stimulated four-photon parametric processes that may exist in monomode optical fibers may also give rise to the primer for subsequent stimulated Raman gain in our case, in addition to self-phase modulation [56-58]. Here we note that there is no fundamental difference between self-phase modulation and the four-photon parametric processes associated with the same real part of the third order nonlinear susceptibility [32, 36].

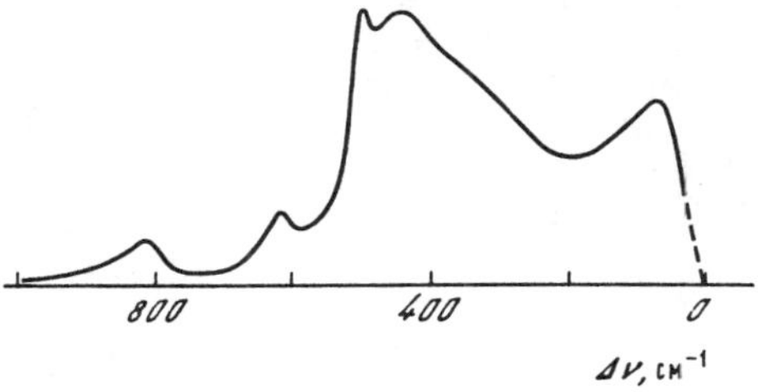

Fig. 6. The spontaneous Raman scattering spectrum in fused vitreous silica glass [67]

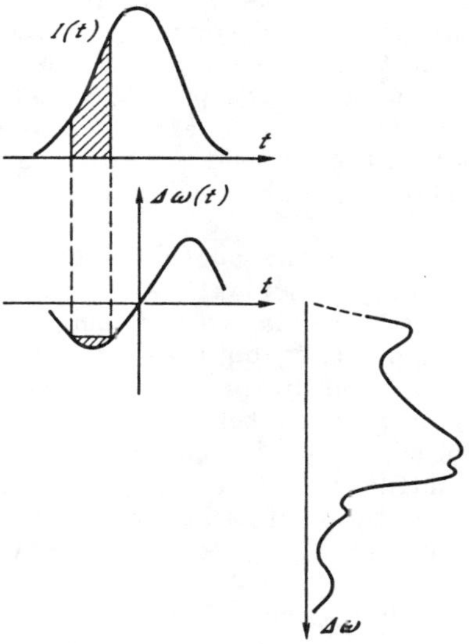

Fig. 7. Formation of primer for stimulated Raman scattering gain from the self-phase modulation effect
The spontaneous Raman scattering spectrum is given on the right.

Calculations from studies [26, 36, 37] also reveal that the formation of the pulse envelope shock wave deriving from the nonlinear delay of the pulse peak with respect to its pedestal will not be evident at the rather long pump pulse duration (30 ps) for our optical fiber length L = 250 m and at launch powers of $P < 1$ kw. A shock wave may form with a narrow, high-intensity peak near the maximum self-steepening range of the pump pulse (see Fig. 4). However a compressed pulse spike contains little pulse energy and therefore such a mechanism cannot explain the significant spectral asymmetry observed in experiment (see Fig. 4d), although this mechanism may play a role in the formation of the primer for subsequent stimulated Raman gain as discussed above.

The stimulated Raman conversion of multisoliton pulses has been examined theoretically [72, 73] which demonstrated the possibility of stimulated Raman conversion of a three-soliton pulse into a single-soliton pulse at the Stokes frequency. The single-soliton pulse accumulates a greater degree of pump energy and therefore has a significantly higher amplitude and, consequently, a shorter duration than the pump pulse. The calculations from studies [72, 73] assume equality of the group velocities of the stimulated Raman and pump pulses, for which, unlike our case, the pump and stimulated Raman pulses will propagate by different transverse modes in a multimode optical fiber to compensate the material dispersion by intermodal dispersion. Moreover such a stimulated Raman transformation of a three-soliton pulse into a single-soliton pulse was inferred from calculations only by assuming a "primer" of specific amplitude at the Stokes frequency at the entrance to the optical fiber. In our case [74, 75] we related the Stokes wing appearing in the spectrum of the exit radiation (see Fig. 4c, d) to the nonlinear Raman conversion of the pump pulse which itself generates a primer at the Stokes frequency as it propagates through a monomode optical fiber and begins a transformation to a Stokes pulse. With a pump pulse power of $P \approx 900$ w and a duration $\tau_0 \approx 30$ ps ($N \approx 100$) the Stokes pulse at the optical fiber exit will be 150 times narrower than the pump pulse at the optical fiber entrance and in this case will contain approximately 50% of the total energy of the exiting radiation.

Since the pulse duration was measured by the correlation technique, it is impossible to establish the waveform and symmetry of the measured pulses at the pump frequency based on the measured autocorrelation function (see Fig. 5a). With efficient stimulated Raman conversion the pump pulse at the exit of the optical fiber will have a rather complex temporal structure. Based on the autocorrelation function measured at the pump frequency (see Fig. 5a) having a pedestal extending from -50 to +50 ps we may conclude with confidence only that the portion of the pump pulse converted as a result of stimulated Raman scattering is not localized within these limits.

Thus, these experiments reveal that the two nonlinear effects of stimulated Raman scattering and self-phase modulation arising in a monomode silica optical fiber from picosecond pulse propagation with negative group velocity dispersion produces powerful femtosecond pulses without the broad pedestal characteristic of the purely soliton compression technique. In this case the femtosecond pulses have a highly stable amplitude, since they are formed in stimulated Raman scattering not from spontaneous noise but rather from a probe wave self-induced by self-phase modulation. In the absence of a high-energy pedestal such a compressed pulse is localized near the 50% energy level of the pump pulse. In our case the power of a 200 fs pulse at ~55 kw is more than 60 times greater than the 30 ps, 900-w pump pulse. The capability for continuous pump pulse tuning over the 1.5-1.65 μm indicates promise for applying such femtosecond pulses to investigations of a variety of ultrafast processes.

II. Stimulated Brillouin Scattering in Monomode Optical Fibers

The interest in stimulated Brillouin scattering in optical fibers derives from several facts, particularly the low critical power of the order of a few milliwatts [76] and the resulting capabilities for obtaining low-threshold phase conjugation [77, 78], conversion of CW radiation into pulsed radiation [79], achieving stimulated Brillouin scattering in a cavity at submillimeter threshold power levels [80]; the development of a stimulated Brillouin scattering-based inertial sensor [81], frequency stabilization of a monofrequency CW laser [82], etc. In coherent fiber-optic links stimulated Brillouin scattering is a harmful effect that limits data transmission range and measures must be taken to suppress this effect [83]. Moreover, the extensive capabilities for varying the length of optical fibers (from a few centimeters to several kilometers) together with the pump power (from a few milliwatts to tens of kilowatts) and the modal composition of the pump and stimulated scattering radiation (from monomode to multimode) makes it possible to achieve several new results. The present study reports results from investigations of the temporal characteristics of stimulated Brillouin scattering in monomode optical fibers.

Optical pump source and instrumentation

These investigations were carried out using the traditional technique for recording backscattering: the experimental configuration is shown in Fig. 8. A CW-pumped, passive Q-switched ($LiF:F_2^-$ crystal [84]) monomode single-frequency $YAG:Nd^{3+}$ laser (λ = 1.06 μm) was used as the pump source. By altering the initial transmission of the shutter it was possible to vary the pump pulse duration from 80 ns to several microseconds. At a pulse duration of 80 ns the pump radiation power was 5 kw. In order to eliminate parasitic reflection from the ends of the fiber, the latter were submerged in glycerin dish 1. The radiation transmitted through the optical fiber and the backscattering were recorded by semiconductor photodiodes 2 whose signals were then injected to oscillograph 3. The recording circuit had a temporal resolution of ~0.7 ns.

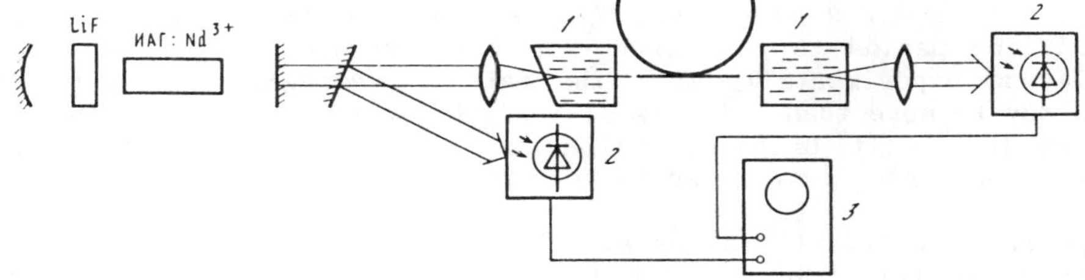

Fig. 8. Experimental set-up

Monomode optical fibers were selected to investigate the temporal characteristics of stimulated Brillouin scattering since monomode optical fiber radiation appears as a wave with a fixed, near-Gaussian transverse field distribution along the entire fiber length. Thus, stimulated Brillouin scattering in a multimode optical fiber is achieved by nonlinear interaction of the spatially-coherent pump and Stokes waves, which guarantees sufficient accuracy in an experiment to investigate the temporal characteristics of stimulated Brillouin scattering and also makes possible direct comparison to the theory developed for plane wave interaction.

The features of stimulated Brillouin scattering in a monomode optical fiber in saturation conditions

The test device in this case was a monomode optical fiber with a core diameter $d_c \approx 6$ μm and a difference of the refractive indices between the core and cladding of $\Delta n = 4.5 \cdot 10^{-3}$. In an $L = 2.9$ m fiber where the round trip period of the medium $T_0 = 2Ln/c = 28$ ns is comparable to the relaxation time of hypersound $T_p \approx 10$ ns and at a pump pulse duration $\tau_H \approx 90$ ns in addition to the stimulated Brillouin scattering generation of a flat pulse whose waveform and duration is identical to that of the pump pulse (Fig. 9a) as observed previously in homogeneous bulk media (see, for example, [85, 86]), we also discovered a periodic sequence of nanosecond pulses appearing on the pedestal whose waveform and duration were identical to that of the pump pulse (Fig. 9b, c) [87]. It has been established experimentally that when the lasing threshold is significantly exceeded in a pump pulse duration range of $\tau_H = 80$–300 ns the repetition period of the pulse peaks (pulses with $\tau_H \approx 1$ ns) is independent of both the power level and duration of the pump pulse. The repetition period of the pulse peaks is dependent on the optical fiber length: at lengths $L =$ 2.9, 1.9, and 1.1 m the repetition period is 56, 53, and 35 ns, respectively.

It has been demonstrated in study [88] that saturation oscillations may occur in the stimulated Brillouin scattering intensity while the period of these oscillations corresponds to the round-trip time through the medium $T_0 = 2Ln/c$. We observed similar oscillations in the stimulated Brillouin scattering effect for an optical fiber length $L = 24$ m and a pump pulse duration $\tau_H = 4$ μs. The repetition period of the nanosecond pulse peaks for three optical fiber lengths (1.1, 1.9, and 2.9 m) exceeds T_0. However, we relate the pulse peak repetition period to the round-trip time through the medium: we established experimentally that the rate of propagation of the pulse peak may be more than twice as slow as the speed of light in the given medium [87]. It is possible that the shortening of the pulse peak repetition period is related to this fact.

It is difficult to explain the generation of narrow nanosecond pulse peaks (Fig. 9b, c) within the framework of existing concepts. In principle when investigating stimulated Brillouin scattering in

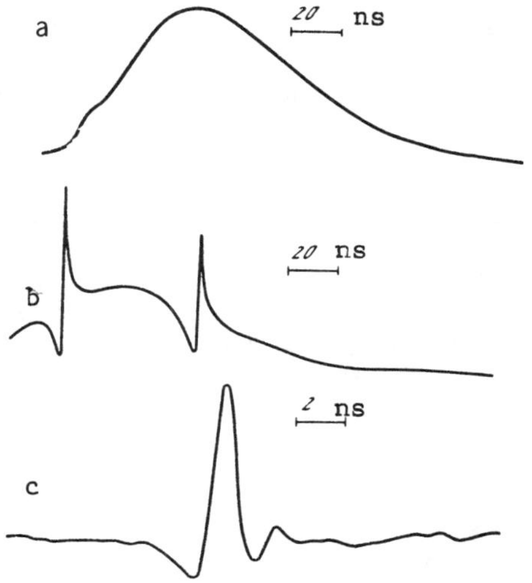

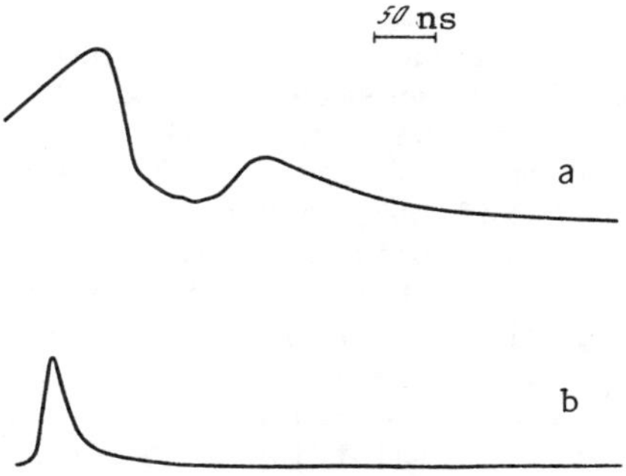

Fig. 9. Oscillograms of stimulated Brillouin scattering in an L = 2.9 m monomode optical fiber for a pump pulse duration τ_H = 90 ns a - plane-wave pulse generation; b - nanosecond pulse peak mode; c- nanosecond pulse peak

Fig. 10. Stimulated Brillouin scattering generation of a single compressed (with respect to the pump pulse) pulse of τ_H = 250 ns (a) and pump radiation transmitted through the optical fiber of length L = 24 m

optical fibers it is necessary to account for acoustic wave diffraction. The diffraction dispersion length $l_д \approx \frac{2\pi}{\lambda_{зв}} a_c^2 \approx 140$ μm (a_c = 3 μm is the radius of the optical fiber core, λ_{3B} = 0.37 μm is the hypersound wavelength: $\lambda_{3B} = \upsilon_{3B}/\Delta\nu$, where $\upsilon_{3B} = 5.9 \cdot 10^5$ cm/s is the speed of sound, $\Delta\nu = 1.6 \cdot 10^{10}$ Hz is the stimulated Brillouin scattering frequency shift [89]) is comparable within an order of magnitude to the free path length of hypersound $l_{3B} \approx y_{3B} T_p \approx 60$ μm. As the hypersound is diffracted its wavefront is distorted, causing the longitudinal component of the wave vector to drop in accordance with the diffraction angle. The change in the longitudinal component of the wave vector of the hypersonic wave will generate a shift in the phases of the pump, Stokes and hypersound waves, which may be because of the anomalies we observed in the temporal structure. Moreover, the transverse field distribution of the acoustic wave which influences the stimulated Brillouin scattering efficiency is determined by the competition between hypersound gain from stimulated scattering (the pump and Stokes waves generating the hypersound attempt to convey their transverse distribution) and its diffraction divergence. The violation of wave synchronism between the waves interacting from stimulated

Brillouin scattering and the attendant appearance of the nanosecond pulse peak changes the acoustic wave gain. This in turn alters the transverse distribution of the hypersound and as a result changes the pump wave/hypersound interaction efficiency. Restoration of the previous scattering conditions may result in generation of a new pulse peak. This is one possible explanation for the periodicity of the nanosecond pulses.

We obtained a Stokes pulse with $\tau_C \approx 16$ ns (Fig. 10b) in another experiment whose purpose was to obtain a single compressed stimulated Brillouin scattering pulse in an $L = 24$ m optical fiber with a pump pulse duration of $\tau_H = 250$ ns [87]. Our estimate of the energy conversion of pump radiation into stimulated Brillouin scattering based on the distortion to the pump pulse transmitted through the optical fiber (Fig. 10a) yields a value of at least 40%. The Stokes emission power significantly exceeds the pump power. This contradicts the notion of stimulated Brillouin scattering generated in a parallel pump beam (see, for example, [90]) which is realized in a monomode optical fiber that maintains a fixed light field configuration along the entire optical fiber length. We attribute the observed compression of the stimulated Brillouin scattering pulse to a possible dependence of the route stimulated scattering gain along the interaction length [90]. Such an effect may be caused by acoustic wave diffraction. The important role of diffraction effects in hypersound have been confirmed by experiments to measure stimulated Brillouin scattering thresholds and for comparing such thresholds in optical fibers with different hypersonic propagation conditions.

Measurement of stimulated Brillouin scattering threshold in fused silica optical fibers with various doping impurities

Both optical and acoustic transverse inhomogeneity is produced in SiO_2-based optical fibers due to the introduction of doping impurities into the core and cladding material. Both waveguide ($\upsilon_{3B}' < \upsilon_{3B}''$), and nonwaveguide ($\upsilon_{3B}' > \upsilon_{3B}''$) hypersonic propagation is possible. Here υ_{3B}' and υ_{3B}'' are the hypersonic velocities in the optical fiber core and cladding, respectively. It should be noted that the significant hypersonic attenuation eliminates the possibility for the stable transverse modal structure characteristic of radiation propagation in an optical fiber. In order to establish the influence of the hypersonic transverse inhomogeneity we measured the stimulated Brillouin scattering gains in monomode optical fibers with different doping impurities.

The stimulated Brillouin scattering gain increment M in which the stimulated scattering power reaches a few percent of the pump power is equal to 16 [91]:

$$M = gIL = g\frac{P}{S_{эф}}L \approx 16. \tag{18}$$

Here g is the stimulated Brillouin scattering gain; I is radiation intensity; P is radiation power; $S_{эф}$ is the effective cross-section of the optical fiber.

We carried out a near field measurement of the intensity distribution in each of the test optical fibers and used these measurements as the basis for calculating $S_{эф}$. By measuring the threshold stimulated Brillouin scattering power we may calculate the gain from (18).

The stimulated Brillouin scattering threshold was measured by exciting the optical fiber with linearly-polarized radiation with a pump pulse duration τ_H = 250 ns. The table gives the primary optical fiber parameters, and the pump power for a polarization azimuth corresponding to minimum threshold and calculation results for the stimulated Brillouin scattering gains with a total measurement error of 20%.

Measurement of stimulated Brillouin scattering gain in monomode optical fibers

Optical Fiber No.	Material		Dif. in refractive indices of core and cladding $\Delta n \cdot 10^{-3}$	Core dia. d_c, μm	Optical fiber length L, m	Effective optical fiber cross-section cm^2	Threshold power W	Gain $g \cdot 10^9$ cm/W
	Cores	Claddings						
1	$SiO_2 + GeO_2$	SiO_2	7	4.3	2.1	2.9	8.7	12.5 ± 0.5
2	$SiO_2 + GeO_2$	SiO_2	4.5	6.0	2.8	4.7	16.7	1.6 ±0.3
3	$SiO_2 + GeO_2$	SiO_2	3.7	6.2	3.1	7.2	10.8	3.4 ±0.7
4	$SiO_2 + GeO_2$	$SiO_2+GeO_2+B_2O_3$	3.1	5.6	3.1	7.2	62.4	0.6 ±0.1
5	$SiO_2 + GeO_2$	$SiO_2+GeO_2+B_2O_3$	3.1	8.8	3.1	6.8	46.8	0.7 ±0.1

The gains calculated by the measured data in optical fibers 1-3 (see table) do not differ significantly. We note that hyperacoustic waveguiding is achieved in optical fibers 1-3 whose cores are doped by GeO_2 while the cladding consists of pure SiO_2 [92]. The gain differential by a factor of 2 in optical fibers 2 and 3 is clearly related to the fact that linear polarization is not conserved in optical fiber 2 with weak linear birefringence [93] which serves to double the stimulated scattering threshold [94]. In optical fiber 2 the stimulated Brillouin scattering threshold is independent of the pump polarization azimuth. On the other hand in optical fiber 3 with strong linear birefringence in which linear polarization is conserved the thresholds for the case where the optical fiber is excited along the linear birefringence axis and at 45° to the axes vary by a factor of near 2 (10.8 and 20 w, respectively). Such a result was also obtained in study [94]. We should note that the measured stimulated Brillouin scattering gain is in satisfactory agreement with the calculated value $g = 5.8 \cdot 10^{-9}$ cm/w [95].

Significantly lower values of g were obtained for optical fibers 4 and 5 which conserve linear polarization. Since the claddings of these optical fibers are doped by B_2O_3 which significantly reduces the acoustic velocity in them [96] hypersonic nonwaveguiding is realized in the optical fiber core. This results in additional hypersonic attenuation in the optical fiber core with stimulated gain due to scattering through the lateral surface, which may be responsible for our measured five-fold reduction in gain.

"Traveling" stimulated Brillouin scattering

As demonstrated in studies [97-99] the noise properties of spontaneous noise amplified by stimulated scattering in saturation conditions are manifest in phase fluctuations of the radiation within the Stokes pulse. Study [97] has observed amplitude fluctuations in spatially-noncoherent stimulated Brillouin scattering radiation in saturation and in phase conjugation slightly over threshold.

Study [00] has investigated both theoretically and experimentally a new method of observing stimulated Brillouin scattering intensity fluctuation distributed along the optical fiber length and attributed to the noise properties of spontaneous scattering. The essence of the method is that if the length of the nonlinear medium is greater than the spatial length of the pump pulse, while the pump power is subthreshold, "traveling" mode stimulated Brillouin scattering is possible. As indicated by theory in this mode the duration of the Stokes pulse exceeds the duration of the pump pulse and may be limited by the length of the medium $\tau_z = 2Ln/c$.

Experiments were performed using a long water cell in a monomode optical fiber in study [100]. Different realizations of the stimulated Brillouin scattering pulse in the "traveling" mode in a L = 200 m monomode optical fiber for a pump pulse duration τ_H = 90 ns are shown in Fig. 11. The duration of the stimulated Brillouin scattering is determined by the length of the medium and is τ_C = 2 μs. An undistorted pump pulse is visible at the edge of each of the oscillograms; this represents reflection off the trailing end of the optical fiber. The mean duration of the separate fluctuation spikes of stimulated Brillouin scattering radiation $\tau_{\Phi П}$ = 25-30 ns correlates with the hypersonic relaxation time in fused silica T_p = 10 ns. An examination of the relationship between the stimulated Brillouin scattering intensity spectrum and its radiation spectrum in the "traveling" mode would be of indisputable interest.

As discussed above the duration of the stimulated Brillouin scattering pulse in the "traveling" mode may be determined by the

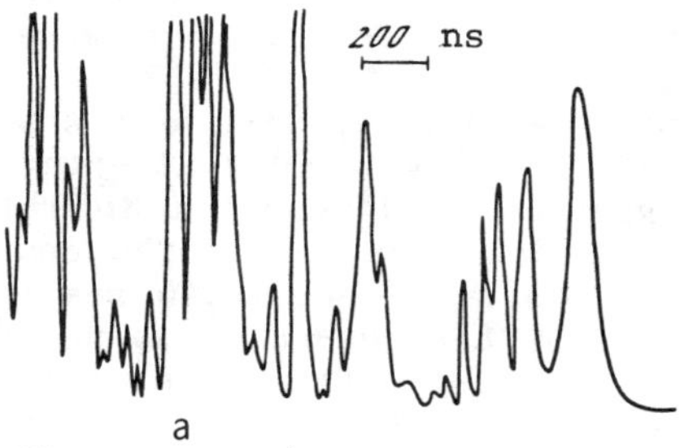

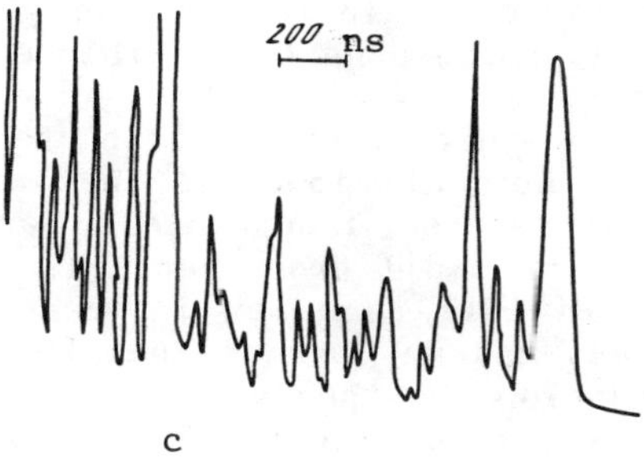

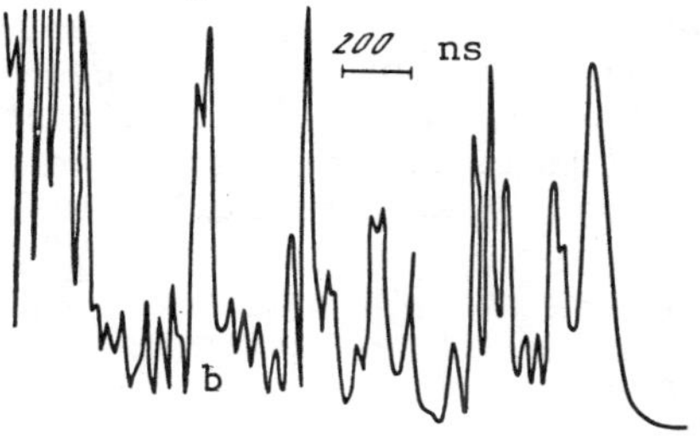

Fig. 11. Oscillograms of stimulated Brillouin scattering in the "traveling" mode in a monomode L = 200 m optical fiber with a pump pulse duration τ_H = 90 ns
a–c – different experimental realizations

interaction length and significantly exceeds the pulse duration. In this case stimulated Brillouin scattering is independently observed from different segments of the optical fiber with a lengthwise resolution corresponding to the duration of the pump pulse: $\Delta l = c\tau_H/2n$. Such a mode makes it possible to obtain information on the characteristics of stimulated Brillouin scattering independently from different segments of a nonlinear medium. For example, in spite of the random nature of the fluctuation structure (Fig. 11) it is possible to identify at least four regions based on the stimulated Brillouin scattering pulse length where the scattering is most intense. Due to its exponential gain the "traveling" stimulated Brillouin scattering mode may be used as a high-sensitivity method of recording inhomogeneities in a medium that apparently are also responsible for the regular variations in stimulated Brillouin scattering intensity along the interaction length.

Conclusion

The present study presents original results obtained from an investigation of the temporal characteristics of stimulated Brillouin scattering in monomode optical fibers. We again emphasize that the fundamental difference between stimulated Brillouin scattering in monomode optical fibers and volumetric media is the fact that the pump and Stokes waves, having an identical transverse field distribution, are spatially-coherent over any length of the medium and with any gain in a monomode optical fiber. It is possible to eliminate the influence of spatially-noncoherent spontaneous emission on the temporal characteristics of stimulated Brillouin scattering in a monomode optical fiber. In addition, the interaction between monomode pump and stimulated Brillouin scattering beams in the optical fiber must adequately correspond to the theoretical model of coupled plane waves.

Another differentiating feature of stimulated Brillouin scattering in optical fibers is the influence on stimulated scattering from the transverse inhomogeneity and hypersonic diffraction in optical fibers of small cross-section. The complex behavior of the acoustic wave in this case may influence the stimulated Brillouin scattering process and may produce new results. In this regard it is necessary to develop the theory for the case of stimulated Brillouin scattering from spontaneous noise and to account for hypersonic diffraction and transverse inhomogeneity. An explanation for the previously unobserved periodic generation of nanosecond pulse peaks and high-energy pulse compression with stimulated Brillouin scattering in a monomode optical fiber [87] will make it possible to establish the influence of these effects on wavefront formation from stimulated Brillouin scattering in multimode optical fibers.

Bibliography

1. "Sverkhkorotkie svetovye impul'sy" [Ultranarrow light pulses] Ed. by C. Shapiro, Moscow: Mir, 1981, 480 p.

2. "Picosecond lasers and applications" Proc. SPIE, Los Angeles, 1982, V. 322, 355 p.

3. Danelyus, R., Dikchus, G., Kabelka, V., et al. "Parametric light excitation in the picosecond range" KVANTOVAYA ELEKTRON., 1977, V. 4, p. 2379-2395.

4. Fomichev, A.A. "New YAG:Nd^{3+} laser lasing states and their possible applications" IZV. AN SSSR. SER. FIZ., 1984, V. 48, p. 2325-2330.

5. Prokhorov, A.M. "Nonlinear effects in optical fibers" Ibid, 1983, V. 47, No. 10, p. 1847-1879.

6. Dianov, Ye.M., Karasik, A.Ya., Prokhorov, A.M., Serkin, V.N. "Nonlinear effects in optical fibers" V. 48, p. 1458-1465.

7. Stolen, R.H. "Nonlinear properties of optical fibers" Optical Fiber Telecommunications, Ed. S.E. Miller, A.G. Chynoweth, NY: Academics, 1979, Chap. 5, p. 125-150.

8. Shimizu, F. "Frequency broadening in liquids by a short-light pulse" PHYS. REV. LETT., 1967, V. 19, p. 1097-1110.

9. Treacy, E.B. "Optical pulse compression with diffraction grating" IEEE J. QUANT. ELECTRON. 1969, V. QE-5, p. 454-458.

10. Fisher, R.A., Kelley, P.L., Gustafson, T.K. "Subpicosecond pulse generation, using the optical Kerr effect" APPL. PHYS. LETT., 1969, V. 14, p. 140-143.

11. Laubereau, A. "External frequency modulation and compression of picosecond pulses" PHYS. LETT. A, 1969, V. 29m, p. 539-540.

12. Lehmberg, R.H., McMahon, J.M. "Compression of 100 ps laser pulses" APPL. PHYS. LETT., 1976, V. 28, p. 204-206.

13. Askar'yan, G.A. "The self-focusing effect" USPEKHI FIZ. NAUK. 1973, V. 111, p. 249-260.

14. Lugovoy, V.N., Prokhorov, A.M. "The theory of high-power laser radiation propagation in a nonlinear medium" Ibid., p. 203-248.

15. Akmanov, S.A., Sukhorukov, A.P., Khokhlov, R.V. "Light self-focusing and diffraction in a nonlinear medium" Ibid., 1967, V. 93, p. 19-70.

16. Ippen, E.P., Shank, C.V., Gustafson, T.K. "Self-phase modulation of picosecond pulses in optical fibers" APPL. PHYS. LETT., 1974, V. 24, p. 190-192.

17. Lugovoy, V.N. "The possibility for frequency scanning of a light pulse field" PIS'MA I ZhETF, 1975, V. 22, No. 8, p. 416-419.

18. Stolen, R.H., Lin, C. "Self-phase modulation in silica optical fibers" PHYS. REV. A, 1978, V. 17, p. 1448-1453.

19. Nicolaus, B., Grischkowsky, D. "90-fs tunable optical pulses obtained by two-stage pulse compression" APPL. PHYS. LETT., 1983, V. 43, p. 228-230.

20. Shank, C.V., Fork, R.L., Yen, R., et al. "Compression of femtosecond-optical pulses" Ibid., 1982, V. 40, p. 761-763.

21. Halbout, J., Grischkowsky, D. "12-fs ultrashort optical pulse compression at a high repetition rate" Ibid., 1984, V. 45, p. 1281-1283.

22. Dianov, Ye.M., Karasik, A.Ya., Mamyshev, P.V., et al. "The effective compression of picosecond pulses from a $YAG:Nd^{3+}$ laser" KVANTOVAYA ELEKTRON., 1984, V. 11, p. 1078-1080.

23. Johnson, A.M., Stolen, R.H., Simpson, W.M. "80-x single-stage compression of frequency-doubled Nondiagonal:yttrium aluminum garnet laser pulses" APPL. PHYS. LETT., 1984, V. 44, p. 729-731.

24. Kafka, J.D., Kolner, B.H., Baer, T., et al. "Compression of pulses from a continuous wave mode-locked Nondiagonal:YAG laser" OPT. LETT., 1984, V. 9, p. 505-506.

25. Ostrovskiy, L.A. "The propagation of light packets and space-time self-focusing in a nonlinear medium" ZhETF, 1966, V. 51, No. 4(10), p. 1189-1194.

26. DeMartini, F., Towns, C.H., Gustafson, T.K., et al. "Self-steepening of light pulses" PHYS. REV., 1967, V. 164, p. 312-323.

27. Ostrovskiy, L.A. "Envelope shock waves" ZhETF, 1968, V. 54, p. 1235-1243.

28. Karpman, V.I., Krushkal', Ye.M. "Modulated waves in nonlinear dispersive media" Ibid, V. 55, p. 530-538.

29. Zel'dovich, B.Ya., Sobel'man, I.I. "The possibility for temporal compression of light pulses in alkali metal vapors" PIS'MA I ZhETF, 1971, V. 13, p. 182-185.

30. Hasegawa, A., Tappert, F. "Transmission of stationary nonlinear optical pulses in dispersive dielectric fibers. I: Anomalous dispersion" APPL. PHYS. LETT., 1973, V. 23, p. 14-144.

31. Vysloujkh, V.A. "Experiments with optical solitons" USPEKHI FIZ. NAUK., 1982, V. 136, p. 519-531.

32. Botineau, J., Stolen, R.H. "Effect of polarization on spectral broadening in optical fibers" J. OPT. SOC. AMER., 1982, V. 72, p. 1592-1596.

33. Owyoung, A., Hellwarth, R.W., George, N. "Intensity-induced changes in optical polarizations in glasses" PHYS. REV. B, 1972, V. 5, p. 628-633.

34. Fisher, R.A., Bishel, W.K. "Numerical studies of the interplay between self-phase modulation and dispersion for intense plane-wave laser pulses" J. APPL. PHYS., 1975, V. 46, p. 4921-4934.

35. Tzoar, N., Jain, M. "Self-phase modulation in long-geometry optical waveguides" PHYS. REV. A, 1981, V. 23, p. 1266-1270.

36. Yang, G., Shen, Y.R. "Spectral broadening of ultrashort pulses in a nonlinear medium" OPT. LETT., 1984, V. 9, p. 510-512.

37. Anderson, D., Lisak, M. "Nonlinear asymmetric self-phase modulation and self-steepening of pulses in long optical waveguides" PHYS. REV. A, 1983, V. 27, p. 1393-1398.

38. Christodoulides, D.N., Joseph, R.I. "Femtosecond solitary waves in optical fibers" ELECTRON. LETT., 1984, V. 20, p. 659-660.

39. Sisakyan, I.N., Shvartsburg, A.B. "The nonlinear dynamics of pico-

second pulses in optical fibers: (a survey)" KVANTOVAYA ELEKTRON., 1984, V. 11, p. 1703-1721.

40. Dianov, Ye.M., Karasik, A.Ya., Mamyshev, P.V., et al. "Direct measurements of monomode optical fiber dispersion in the 1.5-1.4 μm range" PIS'MA I ZTF, 1984, V. 10, p. 518-522.

41. Gloge, D. "Weakly guiding fibers" APPL. OPT., 1971, V. 10, p. 2252-2258.

42. Hasegawa, A., Kodama, Y. "Signal transmission by optical solitons in monomode fiber" PROC. IEEE., 1981, V. 69, p. 1145-115C.

43. Vysloukh, V.A. "Pulse propagation in optical fibers near the dispersion minimum: the role of nonlinearity and higher order dispersion" KVANTOVAYA ELEKTRON., 1983, V. 10, p. 1688-1690.

44. Zakharov, V.Ye., Shabat, A.B. "An exact theory of two-dimensional self-focusing and one-dimensional self-modulation of waves in nonlinear media" ZhETF, 1971, V. 61, p. 118-134.

45. Zakharov, V.Ye., Shabat, A.B. "Soliton interaction in a stable medium" ZhETF, 1973, V. 64, p. 1627-1639.

46. Stolen, R.H., Mollenauer, L.F., Tomlinson, W.J. "Observation of pulse restoration in the soliton period in optical fibers" OPT. LETT., 1983, V. 8, p. 186-188.

47. Satsuma, J., Yajima, N. "Initial value problems of nonlinear Schrodinger equations" PROGR. THEOR. PHYS. SUPPL., 1974, V. 55, p. 284-306.

48. Mollenauer, L.F., Stolen, R.H., Gordon, J.P. "Experimental observation of picosecond pulse narrowing and solitons in optical fibers" PHYS. REV. LETT., 1983, V. 8, p. 289-291.

49. Mollenauer, L.F., Stolen, R.H., Gordon, J.P., et al. "Extreme picosecond pulse narrowing by means of soliton effect in monomode optical fibers" Ibid., 1983, V. 8, p. 289-291.

50. Mollenauer, L.F., Bloom, D.M. "Color-center laser generates picosecond pulses and several watts CW over the 1.24-1.45 m range" OPT. LETT., 1980, V. 45, p. 1095-1098.

51. Basyev, T.T., Lokhnygin, V.D., Mirov, S.B., et al. "Color center generation of quasi-continuous tunable microsecond laser radiation" Materialy IV Vsesoyuz. konf. "Perestraivaemye po chastote lazery [Materials of the 4th All Union Conference "Tunable Lasers"] Novosibirsk, 1983, p. 399-402.

52. Johnson, A.M., Simpson, W.M. "Continuous-wave mode-locked Nd:YAG-pumped lasers" OPT. LETT., 1983, V. 8, p. 554-556.

53. Onishchukov, G.I., Fomichev, A.A., Kholodnykh, A.I. "A picosecond parametric oscillator pumped by a continuously pumped YAG:Nd^{3+} laser" KVANTOVAYA ELEKTRON., 1983, V. 10, p. 1525-1526.

54. Fork, R.L., Shank, C.V., Hirlinann, C., et al. "Femtosecond white-light continuum pulses" OPT. LETT., 1983, V. 8, p. 1-3.

55. Ippen, E.P., Shank, C.V. "Dynamic spectroscopy and subpicosecond pulse compression" APPL. PHYS. LETT., 1975, V. 27, p. 488-490.

56. Lin, C., Reed, W.A., Pearson, A.D., et al. "Phase matching in the minimum chromatic-dispersion region of monomode fiber for stimulated four-photon mixing" OPT. LETT., 1981, V. 6, p. 493-495.

57. Stolen, R.H., Bosch, M.A., Lin, C. "Phase matching in birefringence fibers" Ibid., p. 213-215.

58. Dianov, Ye.M., Zakhidov, E.A., Karasik, A.Ya. "Tuning of four-photon mixing radiation in a birefringent optical fiber" PIS'MA V ZhTF, 1983, V. 9, p. 1455-1459.

59. Dianov, Ye.M., Karasik, A.Ya., Mamyshev, P.V., et al. "The picosecond structure of a stimulated Raman pump pulse in a monomode optical fiber" PIS'MA I ZhETF, 1984, V. 39, p. 564-566.

60. Dianov, Ye.M., Karasik, A.Ya., Mamyshev, P.V., et al. "Spectral filtering of ultranarrow pulses by stimulated Raman scattering in an optical fiber" ZhETF, 1985, V. 88, p. 781-795.

61. Stolen, R.H. "Fiber Raman Lasers" FIBER AND INTEGRAT. OPT., 1980, V. 3, p. 21-51.

62. Grudinin, A.B., Gur'yanov, A.N., Dianov, Ye.M., et al. "The features of stimulated Raman scattered light in SiO_2 + GeO_2 optical fibers" KVANTOVAYA ELEKTRON., 1981, V. 8, p. 2383-2389.

63. Nesterova, Z.V., Aleksandrov, I.V., Polnitskiy, A.A., Sattarov, D.K. "The features of high-power ultranarrow light pulse propagation in multimode optical fibers" PIS'MA I ZhETF, 1981, V. 34, p. 391-395.

64. Stolen, R.H. "Phase-matched stimulated four-photon mixing in silica fiber waveguides" IEEE J. QUANT. ELECTRON., 1975, V. QE-11, p. 100-103.

65. Dianov, Ye.M., Zakhidov, E.A,. Karasik, A.Ya. et al. "Stimulated four-phonon parametric processes in silica optical fibers" PIS'MA I ZhETF, 1981, V. 34, p. 40-41.

66. Dianov, Ye.M., Zakhidov, E.A,. Karasik, A.Ya. et al. "Stimulated

four-phonon nonlinear processes in low-mode silica optical fibers" ZhETF, 1982, V. 83, p. 39-49.

67. Walrafen, G.E., Krishman, P.N. "Model analysis of the Raman spectrum from silica optical fibers" APPL. OPT., 1982, V. 21, p. 359-360.

68. Stolen, R.H., Ippen, E.P. "Raman gain in glass optical waveguides" APPL. PHYS. LETT., 1973, V. 22, p. 276-278.

69. Maier, M., Kaizer, W., Giordmaine, J.A. "Intense light bursts in the stimulated Raman effect" PHYS. REV. LETT., 1966, V. 17, p. 1275-1277.

70. Maier, M., Kaiser, W., Giordmaine, J.A. "Backward stimulated Raman scattering" PHYS. REV., 1969, V. 177, p. 580-599.

71. Washio, K., Inoue, K., Tanigawa, T. "Efficient generation near-IR stimulated light scattering in optical fibers pumped in low-dispersion region at 1.3 m" ELECTRON. LETT., 1980, V. 16, p. 331-333.

72. Bysloukh, V.A., Serkin, V.N. "Generation of high-energy stimulated Raman solitons in optical fibers" PIS'MA I ZhETF, 1983, V. 38, p. 170-172.

73. Bysloukh, V.A., Serkin, V.N. "Nonlinear conversion of solitons in optical fibers" IZV. AN SSSR. SER. FIZ., 1984, V. 48, p. 1777-1781.

74. Dianov, Ye.M., Karasik, A.Ya., Mamyshev, P.V., et al. "100-fold compression of picosecond parametric oscillator pulses in monomode optical fibers in the 1.5-1.65 μm range" PIS'MA I ZhETF, 1984, V. 40, p. 148-150.

75. Dianov, Ye.M., Karasik, A.Ya., Mamyshev, P.V., et al. "Stimulated Raman conversion of multisoliton pulses in silica optical fibers" Ibid., 1985, V. 41, p. 242-244.

76. Cotter, D. "Observation of stimulated Brillouin scattering in low loss silica fiber at 1.3 m" ELECTRON. LETT., 1982, V. 18, p. 495-496.

77. Basiev, T.T., Dianov, Ye.M., Karasik, A.Ya., et al. "Stimulated Brillouin scattering in a multimode silica optical fiber" PIS'MA I ZhETF, 1982, V. 36, p. 85-87.

78. Petrov, M.P., Kuzin, Ye.A. "Stimulated Brillouin scattering in phase conjugation in optical fibers" PIS'MA I ZhETF, 1982, V. 8, p. 729-732.

79. Kawasaki, B.S., Johnson, D.C., Fujii, Y., et al. "Bandwidth-

limited operation of a mode-locked Brillouin parametric oscillator" APPL. PHYS. LETT., 1978, V. 32, p. 429-431.

80. Stokes, L.F., Chodorov, M., Shaw, H.J. "All-fiber stimulated Brillouin ring laser with submilliwatt pump threshold" OPT. LETT., 1982, V. 7, p. 509-511.

81. Thomas, P.J., vanDriel, H.M., Stegeman, G.J. "Possibility of using an optical fiber Brillouin ring laser for inertial sensing" APPL. OPT., 1980, V. 19, p. 1906-1909.

82. Ponikvar, D.R., Ezekiel, S. "Stabilized single-frequency stimulated Brillouin fiber ring laser" OPT. LETT., 1981, V. 6, p. 398-400.

83. Cotter, D. "Suppression of stimulated Brillouin scattering during transmission of high-power narrowband laser light in monomode fiber" ELECTRON. LETT., 1982, V. 18, p. 638-640.

84. Basiev, T.T., Voron'ko, Yu.K., Mirov, S.B., et al. "Efficient passive shutters for $LiF:F_2^-$ crystal neodymium lasers" KVANTOVAYA ELEKTRON., 1982, V. 9, p. 837-839.

85. Heimann, D., Hamilton, D.S., Hellwarth, R.W. "Brillouin scattering measurements in optical glasses" PHYS. REV. B, 1979, V. 19, p. 6583-6592.

86. Maier, M., Rother, W., Kaiser, W. "Time-resolved measurements of stimulated Brillouin scattering" APPL. PHYS. LETT., 1967, V. 10, p. 80-82.

87. Karasik, A.Ya., Luchnikov, A.V. "Generation of nanosecond radiation pulses by stimulated Brillouin scattering in a monomode silica optical fiber" KVANTOVAYA ELEKTRON., 1985, V. 12, p. 1319-1312.

88. Johnson, R.V., Marburger, J.H. "Relaxation oscillations in stimulated Raman and Brillouin scattering" PHYS. REV. A, 1971, V. 4, p. 1175-1182.

89. Basiev, T.T., Dianov, Ye.M., Karasik, A.Ya., et al. "Stimulated Brillouin scattering in silica optical fibers" XI Vsesoyuz. konf. po kogerent. i nelineynoy optike (22-25 noyab. 1982) [The 10th All Union Conference on Coherent and Nonlinear Optics (22-25 Nov, 1982)] Erevan, 1982, V. 2, p. 448-450.

90. Gorbunov, V.A., Papernyy, S.B., Petrov, V.F., Startsev, V.R. "Temporal compression of pulses by stimulated Brillouin scattering in gases" KVANTOVAYA ELEKTRON., 1983, V. 10, p. 1386-1395.

91. Smith, R.G. "Optical power handling capability of low-loss optical

fibers as determined by stimulated Raman and Brillouin scattering" APPL. OPT., 1972, V. 11, p. 2489-2494.

92. Stone, J., Chraplyvy, A.R. "Spontaneous Brillouin noise in a long-distance high-bandwidth optical fiber transmission" ELECTRON. LETT., 1983, V. 19, p. 275-277.

93. Gur'yanov, A.N., Gusovskiy, D.D., Devyatykh, G.G., et al. "The polarization properties of monomode optical fibers with weak birefringence " KVANTOVAYA ELECTRON., 19891, V. 8, p. 2473-2478.

94. Stolen, R.H. "Polarization effects in fiber Raman and Brillouin lasers" IEEE J. QUANT. ELECTRON., 1979, V. QE-15, p. 1157-1160.

95. Ippen, E.P., Stolen, R.H. "Stimulated Brillouin scattering in optical fibers" APPL. PHYS. LETT., 1972, V. 21, p. 539-541.

96. Thomas, P.J., Rowell, N.L., vanDriel, H.M., et al. "Normal acoustic modes and Brillouin scattering in monomode fibers" PHYS. REV. B, 1979, V. 19, p. 4986-4998.

97. Bespalov, V.I., Betin, A.A., Pasmanik, G.A., Shilov, A.A. "Observation of temporal field oscillations in stimulated Brillouin scattering radiation" PIS'MA V ZhETF, 1980, V. 31, p. 668-679.

98. Basil'ev, M.V., Gyulameryan, A.L., Mamaev, A.V., et al. "Recording phase fluctuations in stimulated light scattering" Ibid., p. 673-677.

99. Basov, N.G., Zubarev, I.G., Mironov, A.B., et al. "Phase fluctuations of a Stokes wave from stimulated light scattering" Ibid., p. 685-689.

100. Benenson, Z.M., Bunkin, F.V., Vlasov, D.V., et al. "Stimulated Brillouin scattering in the "traveling" mode" Ibid., 1985, V. 42, p. 164-167.

Fiber-Optic Memories

M.I. Belovolov, Ye.M. Dianov, V.I. Karpov

Abstract: The primary physical principles, primary circuit designs and techniques, and the parameters of fiber-optic memories are examined. It is demonstrated that the memory capacity of such devices may be 10^5–10^6 bits at the data array read/write rate of 1-2 Gb/s with an information storage time of tens of hours. The problems of data storage stability in a fiber-optic dynamic memory under the influence of external temperature changes and mechanical stresses on the optical fiber are examined and a criterion for their tolerable level is established. A description of a fiber-optic dynamic memory for analog signal storage is given.

Introduction

Optical fibers have extensive capabilities in practical application due to the broadband function, low attenuation and noise immunity for signal transmission. These advantages of optical fibers are most fully exploited in monomode optical fibers used in experimental 100-200 km fiber-optic communications links (without repeaters) at a data transmission speed of 1 Gb/s or more [1-5]. Modern graded-index optical fiber manufacturing technology also makes it possible to fabricate broadband long-length links. Study [4] has reported the development of a 37 km fiber-optic communications link with a 400 MHz bandwidth based on a graded-index optical fiber.

However the fiber-optic hardware for high-speed signal transmission and processing including the fiber-optic communications link itself as well as fiber-optic sensors, instruments, and various functional optical fiber components remains incomplete due to the absence of fiber-optic memories that can operate at date transmission speeds of 1 Gb/s or more. The lack of such memories limits the use of high-speed integrated optics analog-to-digital converters for high-speed analog signal processing applications.

Study [6] was the first to propose the application of optical fibers to the fabrication of a new class of memories: fiber-optic dynamic memories. The primary advantages of fiber-optic dynamic memories include high data file read and write speeds compatible in format and transmission speed to high-speed fiber-optic communications links as well as significant memory capacity (10^6 bits or more) and long-term data storage capacity.

The present study which is a continuation of studies [6-9] is devoted to an analysis of the fundamental circuit designs of fiber-optic memories as well as the physical operating principles and applications of fiber-optic dynamic memories. New data are provided on the development and investigation of a prototype analog memory in which a digital fiber-optic memory is used as the base device.

Primary Circuit Designs of Fiber-optic Memories

Optical fibers are used in nanosecond signal delay lines for signal processing in physical experimentation, in radar and nuclear physics [10-11]. A characteristic parameter of a delay line is the product of the bandwidth and the delay which exceeds 10^5 in monomode optical fibers.

The delay of an optical signal in an optical fiber of length L is $t_3 = nL/c$, where n is the group velocity delay factor in the optical fibers; c is the speed of light in the vacuum. We obtain $t_3 \approx$ 1 ms for vitreous silica optical fibers ($n \approx 1.5$) for $L = 200$ km.

Study [10] obtained a 1 ms delay of nanosecond pulses by means of a closed recirculation loop employing a 5.5 km optical fiber and retransmission of pulse amplitude to compensate attenuation. It was noted that noise accumulation from the reception of optical signals with a finite signal-to-noise ratio in emitter-coupled logic devices will cause instabilities in the temporal position of the delayed pulses.

Two nanosecond analog signals were delayed by 0.5 ms in study [11] by means of a recirculation loop employing a 20.4 km optical fiber and an analog repeater. As in the preceding study the number of circulations is limited by noise accumulation.

Strictly speaking passive delay lines and delay lines employing an analog repeater cannot be called memories, since the storage time is determined by the signal-to-noise ratio in the link and no measures are taken to conserve the primary performance characteristics of the signals within a specific tolerable distortion range.

Study [12] was the first to demonstrate the possibility of building an active loop signal delay circuit by establishing conditions for Raman signal gain in an 810 m optical fiber. However noise and fluctuations in the signal and pump emissions also limit storage time.

It is possible to achieve a qualitatively new result if we employ complete amplitude, waveform and temporal regeneration of the data pulses, which makes it possible to store the signals for a time period that far exceeds the signal propagation time through the optical fiber. The memory is then a digital memory.

The primary fiber-optic memory designs and the optical repeater designs are both shown in Fig. 1. The primary fiber-optic dynamic memory circuit is shown in Fig. 1a. Digital data are input through electronic switch *K* to transmitter *П* that modulates laser diode *ЛД*. The data pulses are taken from the exit of the optical fiber, are detected by photodiode *ФД*, are regenerated by regenerator circuit *РГ* and are then sent to the output and to switch *K*, completing the recirculation loop. Switching in *K* is used to erase data and input new data. A local master oscillator may be used to synchronize the operation of the optical repeater circuits and it is not necessary to isolate the clock frequency f_T from the information signal, which simplifies the optical repeater circuit design.

The fiber-optic dynamic memory design is shown in Fig. 1b. All electronic memory components in the optical repeater are housed in a separate optical repeater unit while the optical fiber is wound onto a compact coil. Due to the fact that the memory capacity V is dependent on the optical fiber length which may reach $L \geqslant 100$ km certain features of optical fiber applications in fiber-optic dynamic memories related to optical fiber winding on a compact coil then arise. In this case it is necessary to optimize the waveguide structure and the optical fiber claddings as well as the coil dimensions and material in order to minimize bend and microbend losses in the optical fiber. It is possible to use a number of memories as indicated in Fig. 1c in order to increase memory volume and to store a parallel code. The enormous information capacity of optical fibers is most effectively exploited by employing wavelength-division multiplexing devices *λ-M* (see Fig. 1d). Several independent recirculation loops are implemented at different wavelengths λ_i and information may be input both in parallel and in series.

Thus the primary design principles of the fiber-optic dynamic memories are as follows:

- establishment of a closed data recirculation loop;

- use of an optical repeater with total digital data regeneration;

- use of an optical fiber of sufficient length and a broad bandwidth;

- use of an electronic switch to complete the recirculation loop and for data input-output applications.

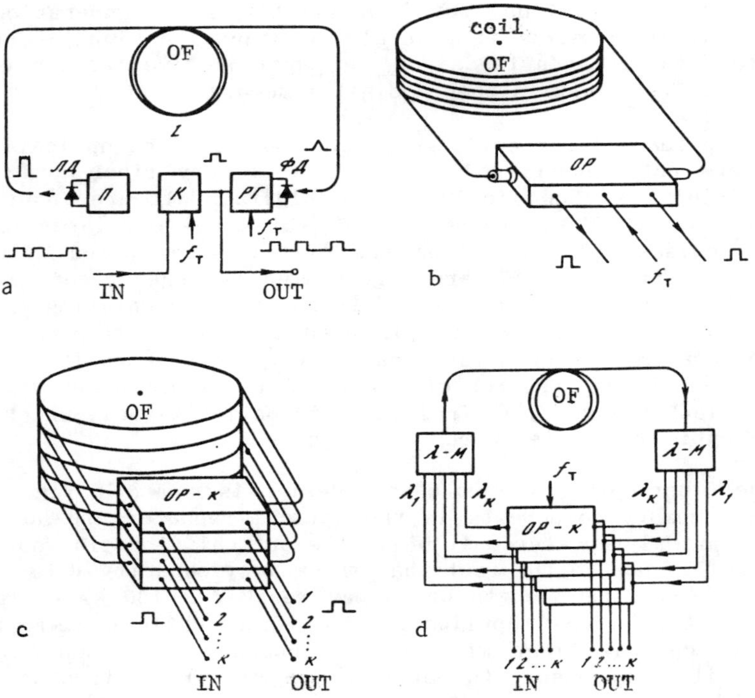

Fig. 1. Schematic diagram of fiber-optic dynamic memory (a), design of fiber-optic dynamic memory (b), fiber-optic dynamic memory for parallel code storage (c) and for spectral channel multiplexing
К - electronic switch; *П* - transmitter; *ЛД* - laser diode; *ФД* - photodiode; *РГ* - repeater; BC - optical fibers; *Вж* - input; *Внж* - output

The parameters of these components will determine such specifications of the fiber-optic dynamic memory as memory capacity, storage time and data read and write time. The fundamental design principles of the memory outlined above require physical substantiation both to establish the fundamental possibility for implementation of a long-term storage mode and to establish the potential capabilities of fiber-optic dynamic memories.

The Physical Operating Principles of Fiber-Optic Dynamic Memories

Normally errors accumulate over time in memories and data are lost. The data storage time is determined by the tolerable level of accumulated error which is a specific value in each concrete case and is to some degree arbitrary. A tolerable accumulated error of 10^{-9} may be established for fiber-optic dynamic memories storing digital data.

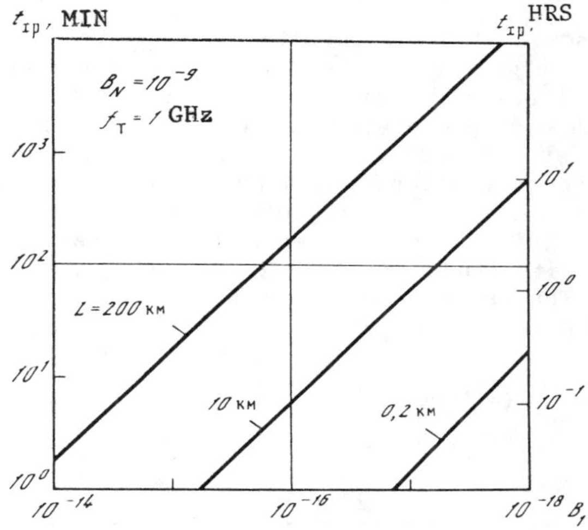

Fig. 2. Data storage time t_{xp} in the fiber-optic dynamic memory plotted as a function of repeater error probability B_1 for various optical fiber lengths L

If the error $B_N = 10^{-9}$ was accumulated in N repeater cycles in the memory, the repeater performance must be such that it provides an error probability for a single repeater cycle equal to $B_1 = B_N/N$. With an optical fiber length L the data storage time in the fiber-optic dynamic memory is determined by the formula [6]

$$t_{xp} = LnN/c, \tag{1}$$

while the memory capacity in bits (the number of pulses circulating in the loop) is determined by the formula

$$V = Lnf_T/c, \tag{2}$$

where f_T is the clock frequency corresponding to the data transmission rate through the optical fiber in the memory.

The data storage time t_{xp} plotted as a function of error probability B_1 is given in Fig. 2 for several optical fiber lengths in a fiber-optic dynamic memory. It is clearly possible to achieve a storage time from a few minutes to tens of hours using rather low error probability values: $B_1 \approx 10^{-14}$–10^{-18}. A more stringent performance requirement of repeaters is one feature of their application to fiber-optic dynamic memories. In long-range communications links employing multiple repeaters an error probability below 10^{-14} is not required, since the error probability 10^{-12}–10^{-14} already makes it possible to employ hundreds of repeaters in a single link. As is indicated by Fig. 2 there may be 10^5–10^9 repeaters in a fiber-optic dynamic memory, thereby dictating such performance requirements on the optical fiber.

We will show that it is possible to achieve an error probability of 10^{-18} without significant loss of range. We will assume that the errors in digital pulse reception result solely from thermal noise in the optical repeater photodetector, while the decision threshold in the repeater is at the half-amplitude point of the data pulses. With these simplifying assumptions which are close to the actual situation in repeaters in the absence of intercharacter interference, we may characterize an information sequence by the parameter $Q = u_c/2u_ш$, where u_c is the amplitude of the signal voltage, while $u_ш$ is the RMS noise voltage. The error probability for the case of a Gaussian noise distribution is given by the expression [13]

$$P_E(Q) = \frac{1}{\sqrt{2\pi}} \int_Q^\infty \exp(-x^2/2)dx. \tag{3}$$

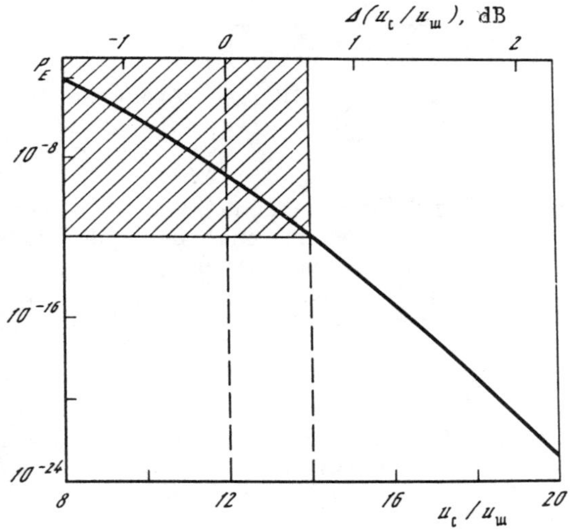

Fig. 3. Repeater error probability P_E plotted as a function of the signal-to-noise ratio $(u_c/u_ш)$ and its variation $\Delta(u_c/u_ш)$ with respect to the point where $P_E = 10^{-9}$

This integral is known as the error function and is tabulated to values of $Q \approx 7$ where $P_E \approx 10^{-12}$. As discussed above there is no practical requirement of $P_E > 10^{-14}$ for communications systems. Fig.

3 shows the error probability P_E plotted against the signal-to-noise ratio $u_C/u_{Ш}$. The hatched region corresponds to the signal-to-noise ratios and error probabilities normally encountered in communications systems.

We have calculated the $P_E(Q)$ relation of interest to us in the range $P_E = 10^{-14}-10^{-23}$ beyond the hatched region in Fig. 3. It is clear that in order to reduce the error probability from 10^{-9} to 10^{-18} it is sufficient to increase the receiver signal-to-noise ratio from 12 to 18, i.e., to raise the pulse energy by only 1.8 dB (the upper scale in Fig. 3), which is possible due to the energy potential of the communications link in the fiber-optic dynamic memory.

These calculations reveal the possibility of achieving a memory mode in a loop optical fiber/optical repeater circuit (see Fig. 1a). A storage time of ~10 h was achieved with a signal-to-noise ratio $u_C/u_{Ш} \approx 30$ with a safety margin in the memory prototype described in study [6], thereby experimentally confirming the calculations and the fundamental possibility of developing a fiber-optic digital memory.

Memory Stability

We will consider external influences (temperature, mechanical stresses, etc.) and optical fiber design on the data storage stability in a fiber-optic dynamic memory.

Calculations and experiments show that a digital pulse sequence may be regenerated with sufficiently low error probability with $u_C/u_{Ш} \geqslant 18$. Normally the clock frequency f_T is detected from the signal for technical implementation of this operation in a regenerator. In the case of a fiber-optic dynamic memory it is possible to use the clock frequency of a local master oscillator which must be sufficiently stable.

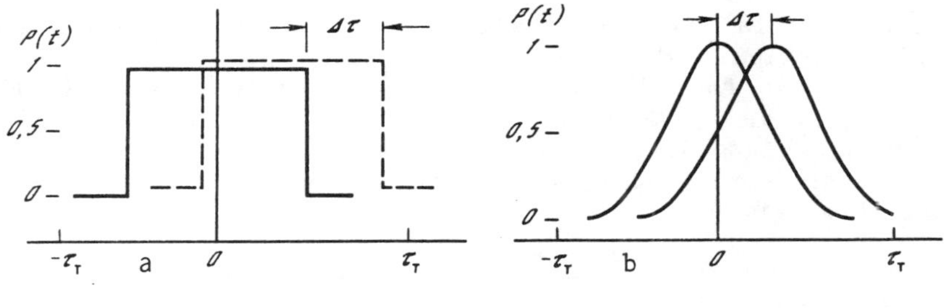

Fig. 4. Tolerable mismatch of the square-wave (a) and Gaussian (b) data pulse waveforms with respect to the origin 0

Clock synchronism must be conserved in regenerating a digital pulse sequence in a recirculation loop, i.e., conserving the ratio of the propagation time τ_L of the light pulses along the optical fiber in the loop to the clock pulse period τ_T: $\tau_L = m\tau_T$, where m is an integer, which assumes a match between the signals and the clock pulses at the zero origins, τ_T, $2\tau_T$, ..., $m\tau_T$, as shown schematically in Fig. 4 for square-wave and Gaussian pulse waveforms. Clock synchronism thus assumes stability of the group propagation time of the light pulses along the optical fiber whose length may extend to $L \approx 100$-200 km.

In actual fiber-optic dynamic memory operating conditions the optical fiber may be influenced by a variety of external actions and the problem of conserving clock synchronism and memory stability requires special attention. Noise in the electronic components of the memory may also produce temporal instability of the data pulses with respect to the clock pulses and criteria for the tolerable noise level must be established.

Fig. 4a shows the tolerable mismatch of $\Delta\tau < 0.5\tau_T$ between a square-wave data pulse and the origin, and this value will not change the signal-to-noise ratio at the decision point. In the case of Gaussian data pulses ($P(t)$ σ $\exp(-t^2/2\sigma_T^2)$, $\sigma_T = 0.25\tau_T$), as shown in Fig. 4b, a mismatch of $\Delta\tau \approx 0.3\tau_T$ will cause the signal-to-noise ratio to drop by a factor of 2 at point zero, which makes it necessary to increase pulse energy by 3 dB to conserve the initial error probability. The condition $\Delta\tau \leqslant 0.3\tau_T$ determines the range of permissible external influences for stable operation of a fiber-optic dynamic memory with an optical signal safety factor of $\approx$3 dB. For f_T 1 GHz we obtain $\Delta\tau \leqslant$ 0.3 ns.

With a change in temperature of ΔT and in mechanical stresses of $\Delta\sigma$ the total change in the group delay $\tau_L = Ln/c$ may be written as

$$\Delta\tau_L = \left(\frac{n}{c}\frac{dL}{dT} + \frac{L}{c}\frac{dn}{dT}\right)\Delta T + \left(\frac{n}{c}\frac{dL}{d\sigma} + \frac{L}{c}\frac{dn}{d\sigma}\right)\Delta\sigma. \tag{4}$$

An analysis of the contribution of the various components in formula (4) to the total value of $\Delta\tau_L$ is given in study [14]. According to this study the total value of the first expression in parentheses for silica glass optical fibers is $d\tau_L/dT \approx 40$ ps$\cdot$km$^{-1}\cdot$K^{-1} for uncoated optical fibers or for fibers with only a primary coating (silicone rubber) and $d\tau_L/dT \approx 200$ ps$\cdot$km$^{-1}\cdot$K^{-1} for optical fibers coated by a nylon coating. According to these data nylon coatings significantly (by a factor of approximately 5) degrade the temperature dependence of the group delay time.

A change in temperature of only 1° will cause a change of $\Delta\tau_L$ = 4 ns in the group delay with an optical fiber length L = 100 km and will result in a loss of high-speed data, while the range of permissible data write speeds in the fiber-optic dynamic memory is limited to ~ 100 Mb/s.

A change in stress $\Delta\sigma$ by ~1 MPa may alter the group delay time by $\Delta\tau_L$ = 5 ns for L = 100 km [14]. These data reveal that the optical fiber cannot be exposed to significant mechanical loading. In the fiber-optic dynamic memory the coil containing the optical fiber must provide fixed tensile stress conditions to the optical fiber and when necessary the coil support material should have a temperature coefficient of linear expansion close to that of silica glass. We should note the positive influence of such a primary coating as silicone rubber, since its application to the optical fiber has only a weak influence on the temperature dependence of the group time compared to an uncoated fiber and reduces optical fiber sensitivity to external mechanical stresses [14].

Thus the following conditions and recommendations should be observed when the memory is synchronized by a local master oscillator, based on the estimates described above for stable memory operation:

1) clock synchronism must be maintained to $\Delta\tau \leqslant 0.3\ \tau_T$, where $\tau_T = 1/f^T$;

2) the optical fiber must be temperature-stabilized to better than -0.1°;

3) nylon et al. coatings should not be used on optical fibers, since they reduce the temperature stability of the group delay time; a silicone rubber primary coating may be employed;

4) temperature-stable optical fibers and coatings should be given preference;

5) the optical fiber should be wound on a coil providing low and stable mechanical stresses and very low additional bend and microbend losses.

These stable operating conditions of a fiber-optic dynamic memory are given for a data write speed of approximately 1 Gb/s and an optical fiber length L = 100 km where this type of memory has significant advantages over other memories. At $f_T \leqslant$ 100 MHz and/or comparatively short optical fiber lengths $L \leqslant$ 10 km the temperature and mechanical stress requirements on clock synchronism are reduced by one to two orders of magnitude.

The memory stability conditions outlined above result from the rather rigid temporal lock between the data pulses and the clock pulses when the memory is synchronized by a local master oscillator of clock frequency f_T. Another possible operational technique is to isolate the clock frequency from the signal. Study [15] investigated a high-speed fiber-optic loop communications link (f_T = 5 GHz, L = 1.5 km) with several repeaters and determined that all repeaters could be synchronized in 1-1.5 s, after which digital data memory could be implemented in such a loop fiber-optic communications link with disparate repeaters. The clock frequency isolation instability was <30 ps; and the study reported no loss of memory stability which evidently may be attributed to the short length of the optical fibers.

Analog Memory

We developed an analog memory prototype which employed, in addition to analog-to-digital and digital-to-analog converters, a digital fiber-optic dynamic memory as the high-speed buffer memory for processing of high-speed analog signals and to investigate the functional relations between the memory components. Drawbacks from using high-speed semiconductor memories for such an application include high power consumption for a comparable memory capacity and complex circuit engineering due to the need to provide parallel high-speed dataflow. After the data are stored in the memory it is then processed by computer at the necessary processing rate. The prototype may also be used as an auxiliary device for an oscillograph in order to observe single signals.

A block diagram of the prototype is given in Fig. 5. The closed recirculation loop consists of transmitter *П*, optical fiber *BC*, photodiode *ФД*, amplifier *Y*, comparator *A*8, logic elements *A*9, *A*7 and local delay line τ_1. The digital information is input to the loop through element A_1 functioning as an "OR" gate. Passing through the optical fiber the pulse sequence is detected by silicon avalanche photodiode *ФД*, is amplified by a high-speed amplifier with a 400 MHz upper bandwidth limit and a gain of 25 dB and is then injected to comparator *A*8 whose operate threshold is set by resistor *R*. Coincidence circuit *A*9 "locks" the temporal position of the pulses to the clock frequency pulses from clock pulse generator *TГ*.

In order to achieve clock synchronism the period of the clock pulse generator must be a multiple of the pulse propagation time through the loop, which is adjusted by the delay τ_1 within the clock pulse generator period by proper selection of the inverters or by altering the optical fiber length in the link (every 20 cm of optical fiber corresponds to a 1 ns delay). The amplitude-, duration, and phase-restored information pulses are then sent to the transmitter through element *A*7.

The first pulse appearing at the output of *A*9 switches RS-flip-flop *A*10, thereby blocking further information input to the recirculation loop. Before the beginning of the write process a pulse input enable signal is injected through *A*6 to the *S* input of this flip-flop.

An analog-to-digital converter (*A*1) converts the analog signal into a digital signal while the parallel code is converted by register *A*2 into a serial code and is input to the memory. The information is stored in the form of eight-bit words in the fiber-optic dynamic memory. The first bit is assigned to label the beginning of the entire data set, while the second bit indicates the beginning of each word (the synchronization bit), while the next six bits are information bits. The operational cycle of the register is divided into three stages: 1 - register clear and analog-to-digital converter digitization of the present analog signal value; 2 - write information from the analog-to-digital converter; 3 - six shift clocks.

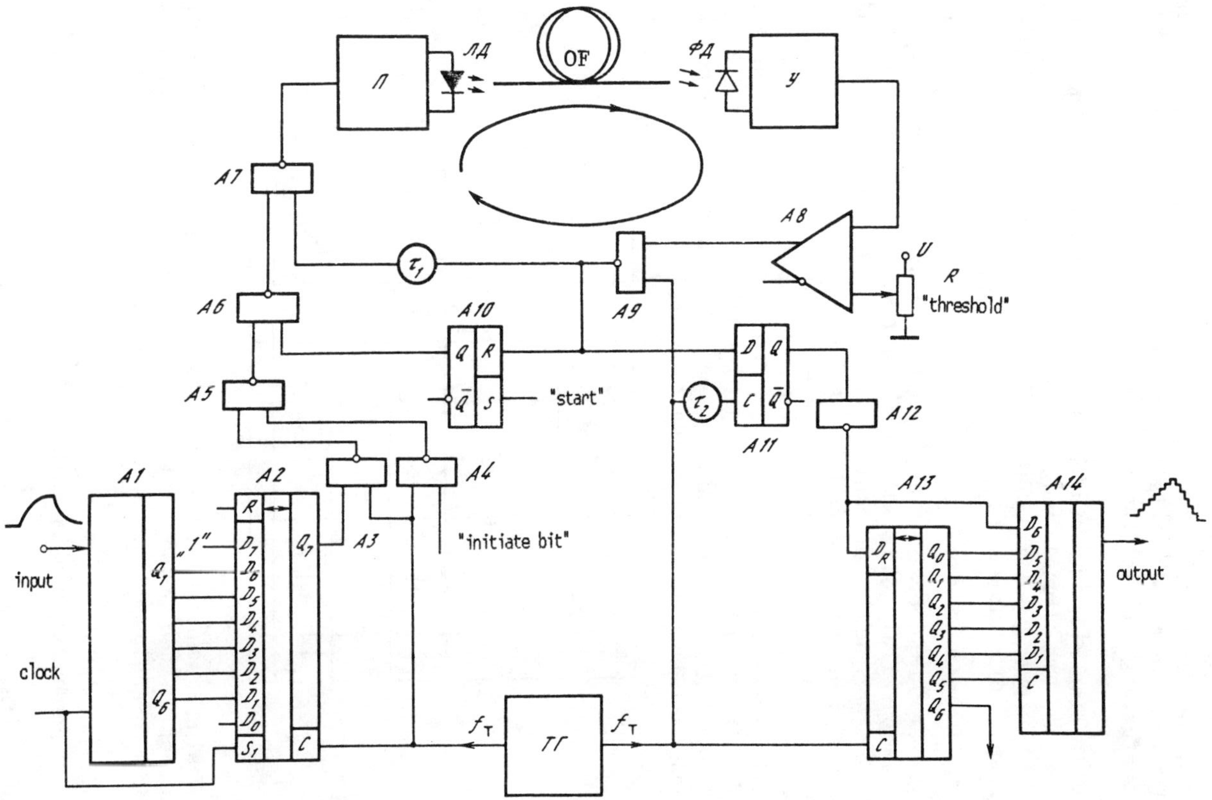

Fig. 5. Schematic diagram of fiber-optic dynamic memory prototype for analog signal storage
П - transmitter; У - amplifier; ТГ - clock pulse generator

In order to simplify regeneration and operation of the laser diode the prototype employs the "return to zero" format. Circuit *A3* converts the signal to this format, and the clock frequency signal is injected to the second input of *A3*. The derived pulse sequence has a relative pulse duration of better than 2. Circuit *A4* generates the array initiate signal which is mixed with the total sequence by circuit *A5* performing an "OR" function.

Д flip-flop *A11* converts the signals back into the "nonreturn to zero" format. The signal sent to input *Д* of this flip-flop is recovered at output *Q* as activated by the leading edge of the sync pulses (input *C*) delayed by $\tau_2 \approx 5$ ns with respect to the clock pulses.

Register *A13* operating as a shift register converts the serial code into a parallel code. The information sequence is sent through invertor *A12* to the "right shift" input D_R of register *A13* and the shift occurs when the clock pulses arrive at input *C*. The appearance of the sync pulse at output Q_5 indicates a full register, and digital-to-analog conversion is activated when this pulse is received. Each first clock clears the register after which it is ready to receive the next word. The recorded analog signal is taken from the output of the digital-to-analog converter (*A14*) in real time and is repeated with a period equal to the data circulation time in the loop.

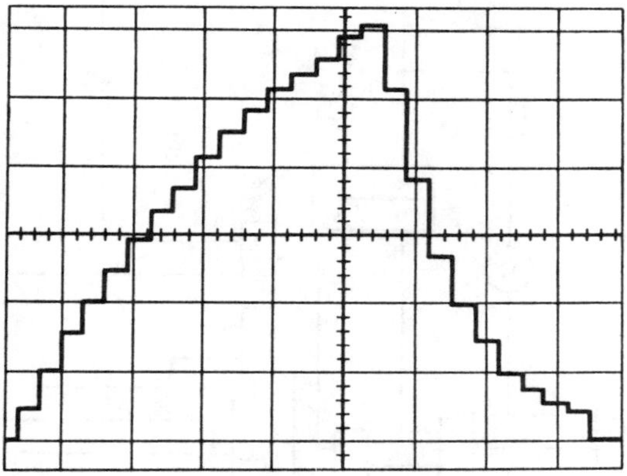

Fig. 6. A single analog signal (a charge-discharge cycle of a capacitor) recorded in the memory and seen on an oscilloscope connected to the digital-to-analog converter output (*A14*)

The regenerator was based on a 597CA2 high-speed comparator and Schottky TTL logic elements. The prototype employed type K1107PV1 analog-to-digital converters and type TsAP-10B digital-to-analog converters. The clock frequency of the pulse sequence is 25 MHz. An optical transmitting module with an AlGaAs-based laser diode radiating at 0.85 μm was used as the transmitter. The fiber-optic link consisted of five SiO_2 + GeO_2/SiO_2-based graded-index optical fiber segments with a total length of 6 km.

With these parameters the device makes it possible to store information on 90 analog signal points with 6-bit coding and 320 ns between samples (sampling frequency of approximately 3 MHz). Since the signal is recorded in real time the duration of the recorded analog signal is determined by the pulse propagation time in the recirculation loop and with an optical fiber length of 6 km this is 30 μs. This memory prototype may be used to process analog signals of significantly greater duration with a sampling period that is an even multiple of the data circulation period in the loop as well as to perform the reverse conversion of digital data from the memory into an analog signal expanded in time; the prototype must have additional electronics for this function.

Fig. 6 shows a single analog signal (the charge-discharge process of a capacitor) recorded in the memory and observed on an oscilloscope at the digital-to-analog converter output (*A*14) (see Fig. 5). The prototype makes it possible to record analog signals with a 64 level amplitude resolution and an upper frequency bound of ~1.5 MHz.

The Technical Parameters and Fields of Application for Fiber-Optic Dynamic Memories

Formulae (1) and (2) indicate that fiber-optic dynamic memories employing optical fibers of maximum length and bandwidth will have the most significant memory volume; the component base of the detectors and radiation sources as well as the optical repeater electronics support such operating parameters. One feature of using optical fibers in a memory compared to a fiber-optic communications link is that with Gaussian information light pulses and a high pulse repetition rate (~1 Gb/s) it is necessary, as discussed above, to provide an optical power safety margin of approximately 2 dB to guarantee a low error probability of the optical detector of ~3 dB and to compensate temperature and other effects. It is necessary to correct the optical fiber length in the memory accounting for such facts.

With a 100 km long optical fiber with a throughput capacity of 4 Gb/s [3] the memory volume of the fiber-optic dynamic memory may be as high as $2 \cdot 10^6$ bits. Such a memory volume is sufficient to store, for example, a complete television picture in a 500 $\times$ 500 element frame employing 8-bit coding. This volume of information may be transferred to another device in 0.5 ms equal to the date propagation time in the loop. In this regard the fiber-optic memory has significant advantages for use as a high-speed buffer or frame memory.

The use of wavelength-division multiplexing in fiber-optic dynamic memories makes it possible to record the same volume of information over an optical fiber length that is *K* times shorter, and the read and write times are reduced by a like amount, where *K* is the number of wavelength-division multiplexed channels.

Fiber-optic dynamic memories are not on a par with modern semiconductor RAMs in terms of accessing time to a single arbitrary data bit, since this time is quite high for optical fiber memories and is equal to the loop recirculation time, and therefore such memories are not suitable for use as a RAM in modern high-speed computers [6].

The broad bandwidth of monomode optical fibers and the operating speed of radiation sources and detectors makes it possible at present to develop communications links with data transmission rates up to 8 Gb/s [5]. In study [15] a data storage mode has already been realized in a loop-type fiber-optic communications link operating at a data transmission rate of 5 Gb/s. Such achievements have highlighted the unique possibilities for employing fiber-optic dynamic memories in conjunction with high-speed integrated optical analog-to-digital converters for high-speed analog signal processing and, to the best of our knowledge, there are no other types of memories aside from fiber-optic dynamic memories that could be used for this application.

We will identify the primary technical fields of application where fiber-optic memories have specific advantages or where they are most convenient.

1. A high-speed computer buffer memory (f_T = 1 GHz, $V \approx 10^6$ bits).

2. High-speed analog signal processing (in conjunction with an analog-to-digital converter, including an integrated optical converter).

3. Television frame scanning memory.

4. Analog and high-speed signal processing with conversion of the time scales of the written and read information.

5. Buffer memory in high-speed fiber-optic communications link.

6. Implementation of a memory mode in loop-type and duplexed fiber-optic communications links [8, 15].

Conclusion

The present analysis of the designs and primary functional aspects fiber-optic dynamic memories demonstrates the fundamental possibility for fabricating such a memory and for achieving its unique parameters representing the most advanced progress of high-speed signal processing and computer hardware. A small reduction in the energy potential and fiber-optic length in the fiber-optic dynamic memory may be used to meet the higher performance requirements on the optical repeater and on the operating parameters of the optical fibers.

An investigation of the problems of memory stability with external influences has made it possible to establish a criterion for their tolerable level as well as to more accurately specify the requirements on optical fibers and their application conditions in fiber-optic dynamic memories. Fiber-optic memories have significant promise for advancement in the direction of assimilation of coherent methods and wavelength-division multiplexing, data transmission by high-speed, ultranarrow pulse packets and the development of waveguiding systems and components in the IR.

In conclusion the authors wish to express their gratitude to Z.S. Nikonova for assistance in the computer calculations.

Bibliography

1. Matsumoto, T., Iwashita, K., Imai, T. "400 Mb/s long-span optical FSK transmission experiment at 1.5 μm" TECH. DIGEST 100C-ECOC'85, Venezia (Italy), 1985, Oct. 1-4, V. 3, p. 31-34.

2. Linke, R.A., Kasper, V.L., Olsson, N.A., et al. "Coherent lightwave transmission over 150 km fiber lengths at 400 Mb/s and 2 Gb/s data rates using DPSK modulation" Ibid., p. 35-38.

3. Korotky, S.K., Eisenstein, G., Gnauck, A.H., et al. "4 Gb/s transmission experiment over 117 km of optical fiber using a $Ti:LiNbO_3$ external modulator" Post deadline Papers OFC/OFS'85, San Diego (Cal.), 1985, Febr. 11-14, Pap. PD1.

4. Appelhans, G., Lackas, W., Lampe, G. "678 Mbaud optical transmission experiment through 37 km of PCVD graded-index fiber" ELECTRON. LETT., 1985, V. 21, No. 7, p. 266-267.

5. Gnauck, A.H., Bowers, J.E., Campbell, J.C. "8-Gb/s transmission over 30 km of optical fiber" TECHN. DIGEST 100C-ECOC'85 Venezia (Italy), 1985, Oct. 1-4, V. 3, p. 44-47.

6. Belovolov, M.I., Golovin, N.I., Golovina, T.N., et al. "Fiber-optic dynamic RAM" KVANTOVAYA ELEKTRON., 1985, V. 12, No. 1, p. 214-216.

7. Belovolov, M.I., Dianov, E.M., Karpov, V.I., et al. "Fiber-optic dynamic memory, First demonstration" Conf. proc. 10-ECOC'84, Stuttgart (FRG), 1984, Sept. 3-6, p. 104-105.

8. Belovolov, M.I., Karpov, V.I. "The stability of a fiber-optic dynamic memory" 2-ya Vsesoyuz. nauch.-tekhn. konf. "Problemy razvitiya radiooptiki" [The 2nd All-Union Scientific and Technical

Conf. "Developmental Problems in Radio Optics"] 4-6 iyunya 1985, Tbilisi: Tez. dokl. No. 1, p. 82-83.

9. Belovolov, M.I., Dianov, Ye.M., Karpov. V.I. "Analog signal processing by means of fiber-optic dynamic memories" Ibid., p. 157-158.

10. Chang, C.T., Lathringer, J.W., Genneten, E.W., et al. "Recirculation of ns pulses through a 5.5 km single-mode fiber optic delay line" TECHN. DIGEST IOOC'83, Tokyo (Japan), 1983, June 27-30, p. 392-393.

11. Wang, C.C., Moeller, R.P., Burns, W.K., Kaminow, I.P. "Fiber-optic recirculating analog delay line" ELECTRON. LETT., 1984, V. 20, No. 12, p. 486-488.

12. Desurvire, E., Digonnet, M., Shaw, H.J. "Raman amplification in a single-mode reentrant fiber loop" TECHN. DIGEST OFC/OFS'85, San Diego (Cal.), 1985, Feb. 11-14, p. 120-121.

13. Unger, G.G. "Opticheskaya svyaz'" [] Moscow: Svyaz', 1979, p. 235-241.

14. Hartog, A.N., Conduit, A.J., Payne, D.N. "Variation of pulse delay with stress and temperature in jacketed and unjacketed optical fibers" OPT. AND QUANT. ELECTRON., 1979, V. 11, p. 265-273.

15. Bergman, L.A., Engineer, S.T. "A synchronous fiber optic ring local area network for multi-gigabit/s mixed traffic communication" Conf. Proc. 10-ECOC'84. Stuttgart (FRG), 1984, Sept. 3-6, p. 272-273.

Elements of Fiber-Optic Communications Links and Their Testing

M.I. Belovolov, A.P. Kryukov, A.V. Kuznetsov, V.Kh. Pencheva

Abstract: This study carries out an experimental investigation of the optical feedback effects arising from the matching of semiconductor lasers to monomode optical fibers. It is demonstrated that such matching elements as focons and microobjectives may be applied in direct detection high-speed fiber-optic communications links, while microobjectives used in coherent fiber-optic communications links must be carefully monitored for their influencing on the lasing linewidth. A method is developed for fabricating fused X- and Y-type low-loss (0.5-1 dB) end-matched directional couplers with a high degree of optical channel isolation (>60 dB). The monomode Y-type directional couplers are used as the basis for developing a new interferometric method of investigating the coherence properties of sources with a high spectral resolution. A simple fiber-optic Michelson interferometer is developed with a discretely variable arm tuning capability and a new visibility recording scheme. With an arm difference of ~1km the interferometer may detect spectral lines less than 0.2 MHz in width which is sufficient for selection and application of radiation sources in coherent fiber-optic communications links.

Introduction

Stringent requirements are imposed on the elements of fiber-optic communications links designed for transmitting high-speed data flows (~1 Gb/s) over distances of 100-300 km without repeaters. This applies in particular to the radiation source whose spectrum must be a monomode spectrum that does not change appreciably when coupled to a fiber-optic link. Particularly stringent conditions are imposed on stable laser operation and spectral bandwidth in coherent fiber-optic communications links (homodyne and heterodyne). There exist data indicating that when focons and microlenses are used to match single-frequency semiconductor lasers to monomode optical fibers, parasitic reflection of radiation off the interface boundaries and Rayleigh backscattering from the optical fiber produce a shift in the lasing frequencies and alters the modal composition of the radiaticn spectrum [1-3]. Therefore it is necessary to investigate optical feedback effects in the various semiconductor laser-monomode optical fiber couplers for both coherent and high-speed direct detection fiber-optic communications links.

The present study investigated certain optical feedback effects arising from the coupling of monomode AlGaAs semiconductor lasers to monomode optical fibers. The parameters and fabrication techniques of low-loss (0.5-1 dB) Y-type couplers are analyzed together with the high optical isolation of convergent channels (60 dB). A new interferometric method of investigating the coherence properties of radiation sources with high spectral resolution is developed based on fiber-optic directional couplers.

Laser-Monomode Optical Fiber Coupling

Two related problems arise when investigating the problems of matching a semiconductor laser to an optical fiber: the first is achieving the maximum possible coupling efficiency η of the semiconductor laser radiation to the optical fiber, and the second problem involves determining the conditions where the optical fiber and its matching device have the least influence on the laser. These issues are normally considered separately in the majority of studies devoted to the efficient coupling of radiation and to optical feedback effects (see, for example, [4-7]).

In calculating coupling efficiency we proceed from the laser radiation field parameters measured in free running lasing without matching devices, while the field distribution is approximated by a Gaussian function. The coupling efficiency η of an elliptical Gaussian beam with characteristic dimensions ω_x and ω_y to an optical fiber with HE_{11} mode dimensions ω_a is expressed by the formula [5]

$$\eta = 4\left[\left(\frac{\omega_x}{\omega_a} + \frac{\omega_a}{\omega_x}\right)\left(\frac{\omega_y}{\omega_a} + \frac{\omega_a}{\omega_y}\right)\right]^{-1}, \tag{1}$$

which indicates that in order to achieve maximum η the laser radiation field parameters must be close to the optical fiber mode dimensions.

The difficulties of matching a semiconductor laser and a monomode optical fiber arise from the fact that neither the characteristic dimensions of the semiconductor laser radiation field nor its angular distribution match the analogous parameters of the HE_{11} mode of the monomode optical fiber. A significant number of studies (see, for example, [4-9] and their references) have been devoted to issues of matching semiconductor laser radiation to a monomode optical fiber. Study [7] has identified the inefficiency of microobjectives, cylindrical and semispherical lenses for coupling monomode optical fibers to stripe semiconductor lasers with respect to domestic prototype radiators, due to significant aberrations. The measured value of η with microobjective matching was 2%.

Study [9] has experimentally investigated the problems of matching radiation from a mesa stripe laser to a monomode optical fiber using bleached lens microobjectives as the matching elements and also provides a description of a simple coupling module. The radiation coupling efficiency in the case of both monomode and multimode lasing spectra for coupling to a monomode optical fiber was 24%. A better microobjective coupling efficiency than reported in study [7] may be achieved with better semiconductor laser and microobjective parameters. Typical values of the constriction parameters at the microobjective focus are $\omega_x = 0.7$ µm and $\omega_y = 0.1$ µm, and when $\omega_a = 2.2$ µm it results in an estimate of $\eta = 0.37$ based on formula (1), at the same time that the experimentally measured coupling efficiency was $\eta_{эксп} = 0.24$. The difference between the estimate and experiment may be attributed to end reflection losses from the monomode optical fiber as well as its imperfections and to errors in the approximation of actual radiation field propagation conditions by Gaussian distributions [5].

The next issue to be considered is the problem of the semiconductor laser radiation spectrum for the case of optical fiber coupling. This problem becomes critical in coherent fiber-optic communications links and is also important for broadband communications systems and fiber-optic systems employing wavelength division multiplexing. When the communications system operates at a wavelength other than the zero material dispersion wavelength in the monomode optical fiber, the mode separation noise of the laser may significantly limit range and data transmission speed [3].

The present study, as a supplement to study [9], investigates optical feedback effects for the case of matching an ILPN-204 semiconductor laser to a monomode optical fiber without matching microobjectives. The coupling efficiency was examined both for flat end optical fiber coupling and for tapered end coupling and spherical microlens coupling (focon coupling).

For the case of semiconductor laser coupling to a flat end optical fiber the coupling efficiencies were $\eta = 0.10$-0.12 for a monomode optical fiber with $\omega_a = 2.2$ µm at $\lambda = 0.86$ µm. The coupling efficiency was improved to $\eta = 0.2$-0.3 by using focons fabricated at the monomode optical fiber end as matching elements. Focon matching, which provided approximately the same coupling efficiency as microobjective coupling, is more compact, although, as will be demonstrated below, it may influence the spectrum of radiation injected to the optical fiber due to optical feedback effects.

The semiconductor laser spectrum is generated by modal interaction (hierarchy), optical limitations on the waveguide structure of the active region, and the selective properties of the Fabry-Perot interferometer formed by the beveled facets of the last crystal. As a rule the modal structure will vary significantly as the pump current, temperature and other factors change. Thus, single-frequency lasing may be detuned by altering the optical feedback level by returning a portion of the radiation power to the cavity. Changes in the radiation spectrum may be explained within the framework of a laser model containing a compound cavity formed by the ends of the optical fiber and the radiating facet of the laser. With a long radiation coherence length the laser spectrum may be influenced by end reflection from the optical fiber and by Rayleigh backscattering [2, 3].

Fig. 1a gives the spectrum of an AlGaAs-based mesa-stripe laser in a free running lasing state at a pump current of 70 ma. The radiation spectrum is near single-frequency (with a spectral resolution ~1 Å) and an intensity ratio of the fundamental to the brightest side mode of 40:1. Fig. 1b shows the radiation spectrum of the same laser when coupled to a flat-end monomode optical fiber 5 μm from the front end of the laser. It is clear that the radiation spectrum of the semiconductor laser was changed significantly, and in addition to the fundamental mode at 8625 Å, two neighboring modes at 8633 and 8636 Å also appeared. Fig. 1c shows the laser spectrum when the optical fiber end is 100 μm from the front end of the laser. The ratio of the fundamental to the most intense lateral mode was 7:1 which indicates optical feedback even at this distance.

Estimates of the optical power level backscattered to the active region of the laser from 100 μm from the flat optical fiber end with 4% Fresnel reflection revealed that this level is approximately 30 dB lower than the radiated power [3], although it does produce significant changes in the radiation spectrum. As noted above, such spectra are completely unsuitable for coherent fiber-optic communications links, while in wideband communications links they may cause laser mode separation noise and alter the degree of channel isolation in wavelength-division multiplexing.

The significant change in the semiconductor laser radiation spectrum from direct coupling of the monomode optical fiber to the flat end indicates a possible change in the near field of the laser radiation and for this reason calculations of η based on measurement data for ω_x and ω_y in free-running lasing will not correspond to the experimentally measured values. The discrepancy between experimentally measured values of ω_x and ω_y and those calculated using data on the far field parameters of the laser radiation support the assumption of a change in the near field of the laser radiation from coupling to the monomode optical fiber.

Fig. 1d shows the laser radiation spectrum for the case of focon coupling. The spectrum is similar to that of a free-running laser (see Fig. 1a). However as indicated by an analysis with a spectral resolution of ~0.5 Å the spectrum expanded from ~1 to ~2 Å, while the neighboring mode suppression was 14:1. Spectral broadening at the 1 Å level is equivalent to detuning by greater than 10 kHz, which is unsuitable for coherent fiber-optic communications links.

Coupling of the single-frequency laser radiation by means of a matching microobjective did not cause a noticeable (in our case accurate to ~0.5 Å) change in the radiation spectrum, since the radia-

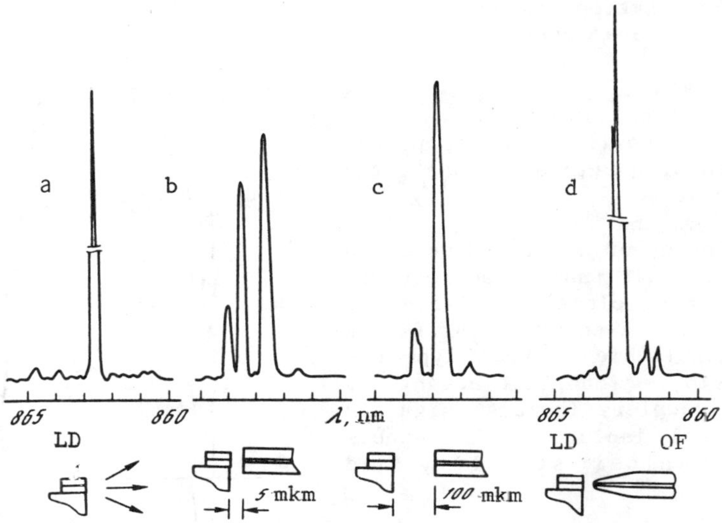

Fig. 1. Radiation spectra for a monomode mesa-stripe AlGaAs laser with different matching conditions to a monomode optical fiber

tion reflected off the fiber end due to low unavoidable detuning as a rule will not impact the active region of the laser, while the Rayleigh backscattering of the optical fiber core material is too weak to cause noticeable changes in the spectrum. Thus, microobjective coupling has greater optical isolation with respect to optical feedback effects than direct coupling. Focon matching with a coupling efficiency of 20-30% may be used in broadband fiber-optic communications links and radiation of ~1 mw may be coupled to a monomode optical fiber.

Microobjective matching may influence the lasing linewidth of a single-frequency laser due to parasitic optical feedback of ≤30 dB, and the suitability of such a method of coupling for coherent fiber-optic communications links requires additional investigation with a spectral resolution on the 10^{-3}-10^{-6} Å level. Special techniques must be employed to investigate the radiation spectra of sources with such a high resolution.

Highly-Directional Fiber-Optic Couplers

Directional couplers are necessary components in a number of fiber-optic communications circuits, test and measurement instrumentation and transducers. The most common requirements on couplers include low attenuation (≤1-2 dB), a high degree of crosstalk attenuation between channels (≥50 dB) which is a measure of the directivity of the coupler, and conservation of the polarization state (in a

Fig. 2. Fabrication stages of X- and Y-type directional couplers

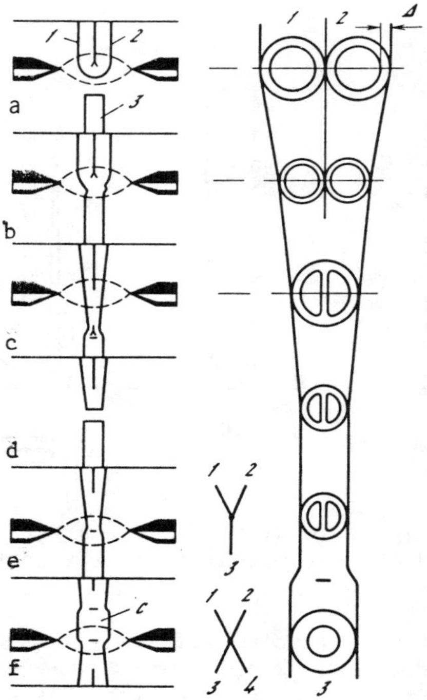

monomode optical fiber) as well as the modal composition of the propagating radiation (in multimode optical fibers).

A survey of studies devoted to couplers reveals that these requirements are most closely satisfied by couplers fabricated by fusing fibers into a monolithic X- or Y-type design [10-15]. Important advantages of fused couplers includes significant optical isolation of channels and mechanical stability and strength.

Our method of fabricating X- and Y-type couplers with high directivity is based on the fusion of optical fibers 1 and 2 (Fig. 2) with partially stripped claddings on an electric arc welder, drawing to a tapered position until a smooth 1-2 mm tapered end is formed and fusion to a single fiber 3 as shown in Fig. 2 for a Y-type coupler. When two tapered ends are fused, an X-type coupler design is obtained. In the case of couplers employing multimode doped core optical fibers, two tapered ends are connected by 200-500 μm optical fiber segments functioning as a mode mixer, as shown in Fig. 2f.

The primary fabrication stages of the couplers are shown in Fig. 2 and characteristic changes in fiber geometry are shown on the right in cross-section in the fusion region: a - melting of the ends of fibers 1 and 2; b - fusion - initial state; c - arc scanning - formation of the tapered end; d - fabrication of the ends; e - final welding of the Y-coupler; f - fabrication of X-coupler with mixer insert C. For all these operations the electrodes as well as the fiber end holders and fasteners were mounted on a three-dimensional support bed for stability accurate to better than 1 μm, which allowed independent electric arc scanning and tapering of the optical fibers. All operations were visually monitored under a microscope and in the last fusion stage (Fig. 2e, f) final tuning was adjusted by transmitting radiation through the coupler.

The coupler parameters are characterized by the power transfer coefficients P_K from channel i to channel j: $\eta_{ij} = P_j/P_i$ or their corresponding attenuations $\alpha_{ij} = 10 \lg P_i/P_j$. A quantitative analysis of the coupler parameters as a function of jacketing thickness Δ which is conserved after etching and the tapering factor $R = d/d_0$ (d_0 and d are the diameters of the fibers before tapering and after, respectively) to achieve minimum losses in Y-type couplers for monomode and multimode optical fibers is given in studies [12, 13]. These investigations proposed geometric-optical models of fused Y-type coupler designs and demonstrated that it is possible to achieve losses of less than 1 dB for a tapering parameter of R = 0.5-0.65 and high optical channel isolation values determined by the Fresnel reflection from the fusion point and Rayleigh scattering at the ends of the initial optical fibers. Study [11] reported a method of fabricating couplers without stripping the jacketing.

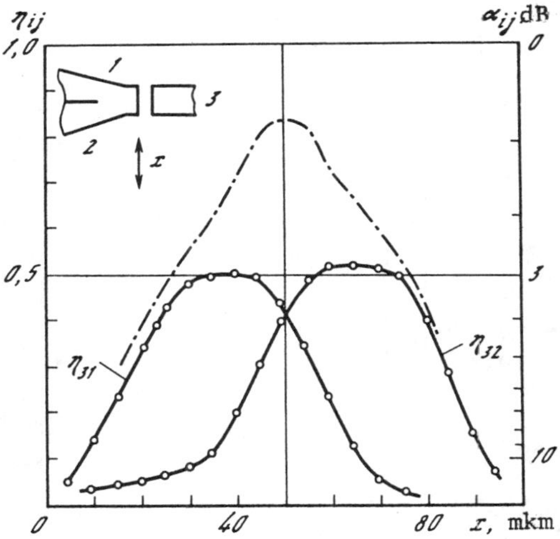

Fig. 3. Power transfer coefficients η_{31} and η_{32} plotted as a function of the radial mismatch X at the fusion site of Y-couplers

Fig. 3 shows the experimental dependencies of the efficiency of optical power transfer η_{31} and η_{32} from single optical fiber 3 to optical fibers 1 and 2 from a transverse mismatch X. The dot and dash curve corresponds to total power transmitted by the coupler and exiting ends 1 and 2. The core diameter of the optical fibers was 50 μm and the jacketing width after exposure of the ends of 1 and 2 was Δ = 10 μm with a tapering factor R = 0.7 and a quasi-parabolic core refractive index profile. Additional losses at the total power peak with equal power levels in channels 1 and 2 amount to 0.7 dB (the average losses for a series of 20 couplers). A mismatch of ±15 μm in Fig. 3 with respect to the origin x = 50 μm makes it possible to attain a 5 to 1 power ratio between channels 1 and 2 with total losses of less than 2 dB.

We should note that if it is necessary to have a significant difference in the radiation power levels from channels 1 and 2 and losses of less than 2 dB it is necessary to employ optical fibers with different core diameters or a single fiber (1 or 2) that is pretapered with the necessary tapering factor. The necessary calculations for obtaining asymmetric low-loss couplers may be carried out analogously to study [13], since the ratio of channel 1 and channel 2 power levels is determined primarily by the overlap of the core areas in the coupler cross-section at the fusion site.

Primary specifications of X- and Y-type couplers

Type of optical fiber, d_c/d_0, μm	Coupler type	Δ, μm	Attenuation constants, dB			λ, μm
			$\alpha_{31} \approx \alpha_{32}$, $\alpha_{41} \approx \alpha_{42}$	$\alpha_{13} \approx \alpha_{23}$, $\alpha_{14} \approx \alpha_{24}$	$\alpha_{12} = \alpha_{21}$, $\alpha_{34} = \alpha_{43}$	
Monomode 7/15	Y	4	4	8	60	0.85
Multimode 50/70	Y	10	3.5	3.5	62	0.85
Silica/polymer 250/250	Y	0	3.6	3.9	50	0.63
					56	0.85
Multimode 50/70	X	10	3.2-3.8	3.2-3.8	57	0.85
Silica/polymer	X	0	3.8	3.8	45	0.63
					50	0.85

Fig. 4. Crosstalk attenuation α_{12} of a Y-type coupler plotted as a function of optical fiber (3) length

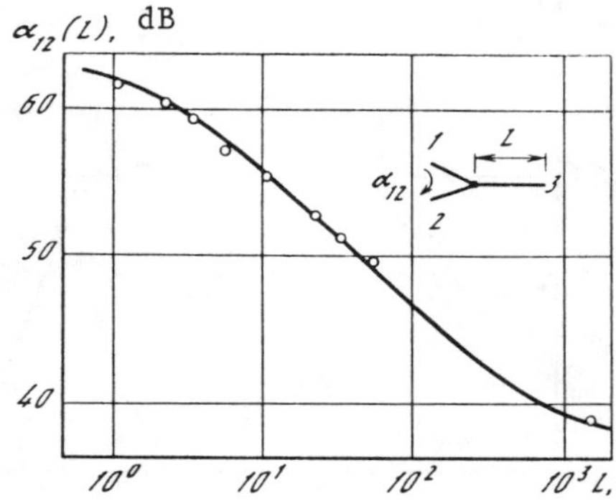

X- and Y-type silica/polymer optical fiber directional couplers (core diameter of 100-300 μm) were fabricated in essentially the same manner as discussed above. Prior to formation of the coupler the protective silicone rubber jacketing was completely removed (Δ = 0), and after fusion of the coupler the cladding was reapplied. The length of the tapered section was 3-5 mm with a tapering factor of R = 0.7 for optimum coupling.

The technique described here was used to fabricate a series of monomode Y-type directional couplers with end matching of single fiber 3 to the tapered end of the two fused and tapered fibers 1 and 2. A detailed analysis of the branching and coupling of the HE_{11}-mode in the coupler design was carried out in study [12] which demonstrated that additional losses may be low (0.51 dB) and the optimum geometric dimensions of such couplers were determined. SiO_2 + GeO_2/SiO_2 monomode optical fibers with a cut-off wavelength of 0.82 μm, a core diameter of 7 μm and an external end diameter of optical fibers 1 and 2 with partially preserved cladding of 15 μm were employed. Additional branching losses were normally 1-3 dB (1 dB in the best samples). The couplers had a high degree of optical directivity (the crosstalk attenuation between channels was $\alpha_{12} = \alpha_{21} \geq 60$ dB), and reduced influence of distributed coupling between fibers 1 and 2 when operating in a coupling mode, and conservation of the polarization state of the propagating radiation.

The table gives the primary parameters of the X- and Y-type couplers measured with all ends carefully exposed in the initial fibers whose length were normally ~1 m. The indices labeling the attenuation constants α_{ij} correspond to the numbering of the coupler ends shown in Fig. 2. It is clear from the table that additional branching and coupling losses in the couplers based on multimode optical fibers amount to 0.5-1 dB. The best of our fabricated monomode optical fiber couplers also had low additional branching losses, and they are promising for investigating backscattering in monomode optical fibers and for sensors and interferometers.

One feature of investigating the parameters of these highly directional couplers from the methodological viewpoint is the need to measure very high crosstalk attenuation between channels such as channels 1 and 2 which requires instrumentation with a significant dynamic range. The luminous flux constants were measured over a 0-80 dB dynamic range by means of a two-channel "optical loss meter" manufactured at the Institute of Electronics of the Bulgarian Academy of Sciences as described in Study [16]. This instrument in conjunction with an additional Y-type coupler used to generate the reference channel were employed to measure the P_i/P_j ratio on a logarithmic scale. ILPN-301 light-emitting diodes, ILPN-204 semiconductor lasers and an LG-52 laser were employed as radiation sources at 0.85 and 0.63 μm.

Fig. 4 shows the crosstalk attenuation α_{12} of channels 1 and 2 plotted as a function of length L of graded-index optical fiber 3 in a Y-coupler having a numerical aperture $NA = 0.16$. The solid line curve represents calculation by formula (1) from study [13], while the dots represent the experimentally-measured values of α_{12}. With an end length $L = 1$ m the contribution to crosstalk attenuation from Rayleigh scattering is vanishingly small and values of $\alpha_{12} = 62$ dB are characteristic of their design. In conjunction with the data from study [13] this represents, to the best of our knowledge, the best experimentally-measured data for crosstalk attenuation in Y-type couplers. An increase in L to approximately 1 km serves to reduce the crosstalk attenuation α_{12} to approximately 39 dB due to the contribution of Rayleigh backscattering from optical fiber 3.

Thus, these couplers have low losses and high directivity, while the monomode couplers also conserve the polarization state of propagating radiation, which makes them promising for application in coherent fiber-optic communications links as well as sensors and monomode fiber-optic test and measurement instrumentation.

The Interferometer and Coherence Testing Procedures for High Spectral Resolution Lasers

Underlying the method of investigating the spectral linewidth of source radiation is, as in Michelson's classical experiments, an analysis of the interference pattern of two light beams of intensities

I_1 and I_2 as a function of their delay τ with respect to one another or a corresponding path difference ΔL. The visibility of the interference bands is determined by the expression [17]

$$V = \frac{E_{max} - E_{min}}{E_{max} + E_{min}} = \frac{2\sqrt{I_1 I_2}}{I_1 + I_2}\gamma(\tau), \tag{2}$$

where E_{max} and E_{min} represent the maximum and minimum illumination of the interference bands near the selected point in the radiation field; $\gamma(\tau)$ is the coherence factor. If the polarizations of the interfering rays coincide and $I_1 = I_2$, the visibility V will qualitatively coincide with the coherence factor $\gamma(\tau)$ of the test radiation.

It may be demonstrated that with a Lorentz spectral radiation line of half-width $\Delta\nu$ the coherence factor is given by the expression

$$\gamma(\tau) = e^{-\tau/\tau_K}, \tag{3}$$

where $\tau_K = 1/\Delta\nu$ is the coherence time determining the spectral linewidth $\Delta\nu$ (or in wavelengths $\Delta\lambda = \lambda^2\Delta\nu/c$) and the coherence length $L_K = \lambda^2/\Delta\lambda = \Delta\nu$.

In coherent fiber-optic communications links the radiation sources must have narrow and stable spectral lines. In order to transmit information at approximately 140 Mb/s a spectral linewidth of $\Delta\nu \approx 100$ kHz is required for phase modulation and $\Delta\nu \cong 1$ MHz for frequency modulation, which corresponds to a spectral linewidth in wavelengths of $\Delta\nu = 10^{-4}$–10^{-5} Å ($\lambda \approx 1$ μm) and coherence lengths of $L_K = 0.1$–1 km [18, 19].

Today heterodyne and homodyne techniques exist for investigating spectral lines up to 0.5 kHz in width; these have been used successfully to investigate the coherence properties of sources for high-resolution spectroscopy, for use as frequency standards and for optical communications [20-22]. From the technical viewpoint these are not simple devices; they require special equipment and precision optical tuning. Scanning Fabry-Perot interferometers, including fiber-optic interferometers, make it possible to achieve a spectral resolution on the order of ≈100 MHz, although they are quite convenient for express spectral measurements of a radiation source [22, 23].

In principle the Michelson interferometer makes it possible to measure the waveform and spectral linewidth of the radiation sources over a broad range, although until recently it was a very difficult practical problem to build an interferometer with an arm difference of ≈0.1–1 km or more. Study [24] for the case of a path difference of $\Delta L = 1$ km estimated the lasing linewidth from a gallium arsenide laser at nitrogen temperatures at less than 150 kHz, while the estimates from study [25] with $\Delta L = 4.15$ km yielded estimates of $\Delta\nu \approx 12.5$ kHz. These results illustrate the significant possibilities for using Michelson interferometers based on monomode optical fibers to produce

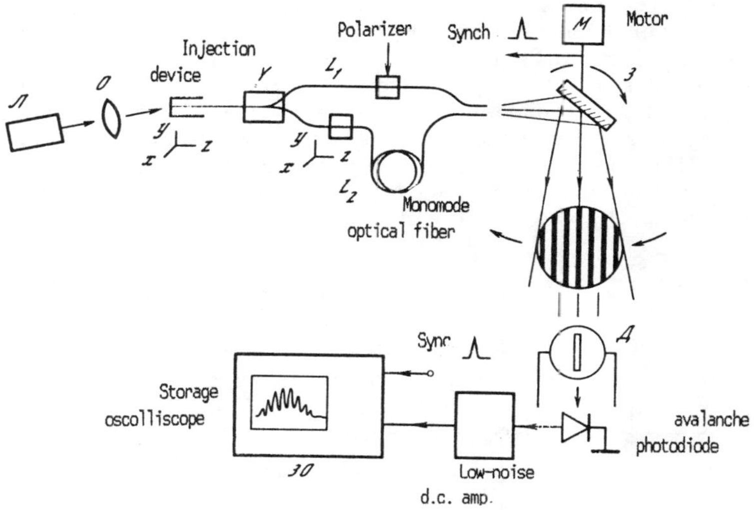

Fig. 5. Block diagram of an experimental assembly containing a fiber-optic Michelson interferometer

an upper estimate of the lasing linewidth of radiation sources for coherent fiber-optic communications links. However in spite of their simplicity, fiber-optic Michelson interferometers have not yet been widely used for this application due to significant radiation losses from the diaphragming of the interference pattern during the recording process and the sensitivity of the results to parasitic reflection from the discrete elements of the optical scheme. Study [26] employed a Mach-Zehnder interferometer in an optical scheme consisting of three X-type couplers used to investigate the linewidth narrowing effect from a semiconductor laser due to Rayleigh backscattering from the adjacent 1 km monomode optical fiber. A linewidth of $\Delta\nu \leqslant 1$ kHz was recorded in this study for a path difference of 40 km for a semiconductor laser operating at near 1.3 μm.

In the present study we first designed an optical scheme for the Michelson interferometer based entirely on a monomode optical fiber, using a single highly directional Y-type coupler as the branching device and a new scheme for recording the visibility of the interference pattern.

A block diagram of the experimental set-up is shown in Fig. 5.

The radiation from laser *Л* was injected to the short (~1 mm) end of the Y-coupler by means of bleached objective *O* and injection device *YB* (the optical fiber connection point to the three-dimensional support frame *x*, *y*, *z*) at a small angle to the end plane in order to avoid parasitic reflection from the end back to the laser. The necessary path difference $\Delta L = L_2 - L_1$ was established by connecting to one of the coupler arms an optical fiber of length L_2 which was discretely tuned in the interferometer, after which the free ends of the coupler were mounted in parallel and their ends were equalized. The beams with a characteristic Gaussian intensity distribution then merge in the far field and interference bands appear that are later analyzed by the visibility recording scheme. The external diameter of the monomode optical fiber (~120 μm) determines the distance between the cores and determines the specific length of the repetition period of the interference bands in the far field.

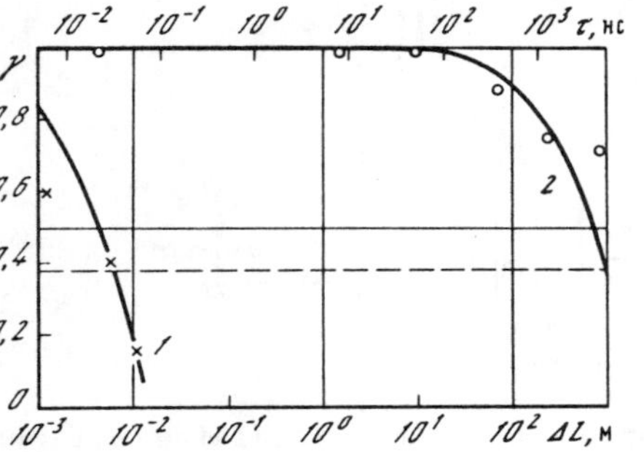

Fig. 6. The coherence factor γ of radiation from several lasers plotted against the path difference ΔL and delay τ
1) $\Delta\lambda \approx 0.8$ Å ($\lambda = 0.85$ μm), $L_k = 9$ mm;
2) $\Delta\lambda \geqslant 3 \cdot 10^{-6}$ Å ($\lambda = 0.63$ μm), $L_k = 1.5 \cdot 10^3$ m

Slit diaphragm *Д* (slit width 100 μm, length 5 mm) was oriented along the interference bands and the radiation then impinged the germanium or silicon avalanche photodiode *ЛФД*. The slit diaphragm increased the signal by virtually two orders of magnitude compared to a circular diaphragm of ~100 μm in diameter which is needed for photometry of the interference bands. The signal is taken from the avalanche photodiode and is passed through a low-noise d.c. amplifier (*УПТ*) to a storage oscilloscope (*ЗО*) for visual monitoring and visibility *V* measurements. Mirror *З* was rotated by motor *M* at 20 Hz for interference pattern scanning and to record oscillations in band illumination from which E_{max} and E_{min} were determined.

The visibility was measured over an oscillation period of ≈30 μs. A pattern stability of ~1 ms is sufficient to record visibility. Radiation frequency fluctuations were then recorded by averaging over ≈30 μs. The maximum visibility for the given path difference ΔL was established by the polarization adjuster *УПП* by rotating the polarization plane of the radiation in the L_2 channel. We should note that the small differential between I_1 and I_2 has no significant influence on *V*. Thus a difference of 20 and 50% would result in an error of 1 and 6%, respectively, in determining *V* according to formula (2). Fig. 6 shows the experimental plots of the coherence factor of the radiation from several lasers versus the path difference of the interfering beams $\Delta L = L_2 - L_1$ or the delay τ. The upper curve was obtained for He-Ne laser radiation (LG-209) which we deliberately did not stabilize and which had sufficient monochromaticity for testing our method of measuring the $\gamma(\tau)$ relation. It is clear that the coherence factor drops to a value of $\gamma = 0.4$ with a path difference of approximately 1 km (in a monomode optical fiber). The solid line curve was drafted by formula (3) for a coherence time τ_K - 5 μs. It follows from these data that the spectral width of the laser radiation does not exceed $\Delta\nu$ = 200 kHz. This is not a record result for an He-Ne laser, although it illustrates the capabilities of the interferometer and the visibility recording technique for investigating radiation sources with a spectral linewidth less than $3 \cdot 10^{-6}$ Å.

Fig. 6 shows the experimental values of the coherence factor obtained for a "single-frequency" semiconductor laser. The coherence factor at $1/e$ determines the coherence length and time of the radiation from this laser from which the spectral linewidth is $\Delta\lambda \approx 0.8$ Å (or $\Delta\nu$ = 30 GHz) and this is in agreement with the value measured by monochromator MDR-4.

The spectral resolution of the interferometer may be significantly improved by increasing the fiber length L_2. However when investigating lasers specifically for coherent fiber-optic communications links it is sufficient to use an arm difference of 1 km which makes it possible to acquire reliable measurements of the spectral linewidth at 0.1 MHz ($\Delta\lambda \approx 10^{-5}$–$10^{-6}$ Å) in the 0.8–1.6 μm wavelength range that is sufficient for virtually any types of modulation. The input end of the Y-coupler in the monomode optical fiber interferometer is virtually an optical probe which may also be used to examine the spatial coherency of the light fields of the sources.

Conclusion

The present study experimentally investigated the optical feedback effects from matching monomode semiconductor lasers to a monomode optical fiber. The experiments revealed that it is possible to use such matching elements as focons and microobjectives in high-speed fiber-optic communications links, while it is advisable to employ microobjectives in coherent communications systems while carefully monitoring their influence on the lasing linewidth of the semiconductor laser with a spectral resolution of better than 10^{-4} Å.

A method was developed for fabricating low-loss fused X- and Y-type directional couplers employing end optical fiber matching characterized by a high degree of optical isolation of the communications channels (greater than 60 dB). In addition the monomode couplers conserve the polarization of the propagating radiation.

A simple fiber-optic Michelson interferometer with a discretely variable arm difference and a new visibility recording scheme that makes it possible to investigate the dependence of the coherence factor of the radiation on the delay time of the interfering beams was developed based on the Y-type monomode couplers employing end matching. With an arm difference of 1 km the interferometer makes it possible to detect spectral lines of width $\Delta\nu \leqslant 0.2$ MHz ($\Delta\lambda \leqslant 10^{-5}$ Å) which is sufficient for selection and application of radiation sources in coherent fiber-optic communications links.

In conclusion the authors wish to express their gratitude to Ye.M. Dianov, A.N. Gur'yanov, D.D. Gusovskiy, and V.F. Khopin for their assistance in the study as well as useful commentary.

Bibliography

1. Vodhanel, R.S., Ko, J.S. "Reflection-induced frequency shift in single-mode laser diodes coupled to optical fibers" Techn. Digest Conf. Opt. Fiber Commun. and Opt. Fiber Sensors. San Diego (USA), 1985, Feb. 11-14, p. 60-63.

2. Wenke, G., Elze, G. "Investigation of optical feedback effects on laser diodes in broadband optical transmission systems" J. OPT. COMMUNS., 1981, V. 2, No. 4, p. 128-133.

3. Peterman, K., Arnold, G. "Noise and distortion characteristics of semiconductor lasers in optical fiber communication systems" IEEE J. QUANT. ELECTRON., 1982, V. QE-18, No. 4, p. 543-554.

4. Khoe, G.D., Kock, H.G., Kuppers, D., et al. "Progress in monomode optical fiber interconnection devices" J. LIGHTWAVE TECHNOL., 1984, V. LT-2, No. 3, p. 217-227.

5. Ramer, O.G. "Single-mode fiber-to-channel waveguide coupling" J. OPT. COMMUNS., 1981, V. 2, No. 4, p. 122-127.

6. Ayunts, Yu.Kh., Belevolov, M.I., Dianov, Ye.M., Tsibulya, A.B. "Optical radiation coupling to elliptical optical fibers" KVANTOVAYA ELEKTRON., 1983, V. 10, No. 12, p. 2433-2443.

7. Ayunts, Yu,Kh., Belovolov, M.I., "Monomode optical fiber matching to semiconductor lasers" Ibid., 1982, V. 9, No. 11, p. 2197-2203.

8. Ayunts, Yu.Kh., Belovolov, M.I., Dianov, Ye.M., et al. "A device for coupling monomode optical fiber to channel lasers" OFT.-MEKH. PROM-ST', 1982, No. 10, p. 58-59.

9. Belovolov, M.I., Dianov, Ye.M., Kryukov, A.P. "An efficient coupling module for coupling the ILPN-204 industrial semiconductor laser to a monomode optical fiber" ELEKTROSVYAZ', 1986, No. 8, p. 51-53.

10. Judy, A.F., Mathis, T.D. "Welded optical fiber signal splitter" Techn. Digest Topic. Meeting Opt. Fiber Commun. Wash. (DISPERSION CURVE), 1979, Mar. 6-8, p. 84-86.

11. Belovolov, M.I., Dianov, Ye.M., Luchnikov, A.V., Prokhorov, A.M. "Low-loss fiber-optic directional couplers" KVANTOVAYA ELEKTRON., 1980, V. 7, No. 7, p. 1578-1580.

12. Belovolov, M.I., Gur'yanov, A.N., Gusovskiy, D.D., et al "Low-loss

directional couplers based on monomode optical fibers" Ibid., 1985, V. 12, No. 9, p. 1873-1880.

13. Belovolov, M.I., Kebedzhiev, A.G., Kuznetsov, A.V. "Highly-directional fiber-optic Y-type low-loss couplers for duplex fiber-optic communications links" ELEKTROSVYAZ', 1985, No. 2, p. 35-39.

14. Emkey, W.L., "Nontapered fused-fiber couplers" TECHN. DIGEST OFC/OFS'85, San Diego (USA), 1985, Feb. 11-14, p. 108-109.

15. Yataki, M.S., Varnham, M.P., Payne, D.N. "Fabrication and properties of very-long fused-taper couplers" Ibid., p. 108-110.

16. Drazhev, M.N., Kebedzhiev, A.G. "A logarithmic picoammeter" PTE, 1980, No. 6, p. 76-78.

17. Dandsberg, G.S. "Optika" [Optics], Moscow: Nauka, 1976, p. 91-112.

18. Yamamoto, Y., Kimura, T. "Coherent optical fiber transmission systems" J. QUANT. ELECTRON. 1981, V. QE-17, No. 6, p. 919-934.

19. Okoshi, T., Kikuchi, K. "Heterodyne-type optical fiber communications" J. OPT. COMMUNS., 1981, V. 2, No. 3, p. 82-88.

20. Okosho, T., Kokushi, K., Nakayama, M. "Novel method for high resolution measurement of laser output spectrum" ELECTRON. LETT., 1980, V. 16, p. 630-631.

21. Akul'shin, A.M., Basov, N.G., Velichanskiy, B.L, et al. "Heterodyne measurements of the lasing linewidth of injection lasers with beat frequency stabilization" KVANTOVAYA ELEKTRON., 1983, V. 10, No. 8, p. 1527-1529.

22. Zubrov, A.S. "A tunable narrow lasing linewidth injection laser" Dissertation for candidate of physics and mathematics 01.04.03, Moscow: FIAN, 1985.

23. Stone, J. "Optical-fiber Fabry-Perot interferometer with finesse of 300" ELECTRON. LETT., 1985, V. 21, No. 11, p. 504-505.

24. Ahearn, W.E., Crowe, J.W. "Linewidth measurements of CW gallium arsenide lasers at 77 K" IEEE J. Quant. Electron, 1966, V. QE-2, No. 9, p. 597-602.

25. Machida, S., Kawana, A., Ishibara, K., Tsuchiya, H. "Interference of an AlGaAs laser diode using 4.15 km single-mode fiber cable" Ibid, 1979, V. QE-15, No. 7, p. 155-157.

26. Epworth, R.E., Smith, D.F., Wright, S. "A practical 1.3 μm semiconductor source with significantly better short term coherence than a gas laser" Conf. Proc. 10th Europ. Conf. Opt. Commun. Stuttgart (FRG), 1984, Sept. 3-6, p. 132-133.

Data Transmission by Spectral- [Wavelength-] Division Multiplexing

A.A. Kuznetsov

Abstract: A series of optical investigations of spectral division channel multiplexing in optical fibers is carried out. Channel multiplexers/demultiplexers based on minispectrometers, microoptics (a focusing rod), integrated optical devices (a plane waveguide and an optical waveguiding glass plate), distributed coupling between monomode optical fibers and based on a pair of superconductors with different bandgaps are proposed and implemented. The applicability of the various test devices is determined in experimental fiber-optic spectral-division multiplexed communications links, specifically a multichannel duplex link in the 1.3 μm range employing a single 5 km optical fiber.

Introduction

The problem of an optical analog to radio frequency-division multiplexing arose in the transition from RF communications to optical communications. The possibility for simultaneous transmission of several dataflows on different optical carriers (beams of different frequency) on a single optical circuit has already been considered for atmospheric laser communications systems and for closed optical systems (with pipeline propagation of optical rays). However this field became practical only with the development of highly-transparent optical fibers. The possibility of "multicolor" (multichannel) transmission on a single glass fiber may be found in the daily observance of a multicolored world through window glass and from simple white light transmission experiments on an optical fiber.

The spectral range in which it is possible to transmit signals at different frequencies on an optical fiber is determined by the spectral window of the fiber or its optical bandwidth (we proposed the latter term in study [1]). In some optical fibers this spectral interval is 1000 nm (0.8-1.8 μm), corresponding to an ultrabroad optical bandwidth of ~$2 \cdot 10^5$ GHz.

The first significant study devoted to optical frequency-division multiplexing in optical fibers was Tomlinson's study [2] that appeared in 1977. Since that time the foreign literature has come to employ the term that translates as "wavelength-division multiplexing" to describe this multiplexing technique (as distinct from frequency-division multiplexing in the RF range). However, frequency and wavelength are unambiguously related characteristics of the same electromagnetic signal and hence the terms "frequency-division multiplexing" and "wavelength-division multiplexing" represent a tautology. For this reason we proposed the term "spectral-division multiplexing" in studies [1, 3] to describe the case of simultaneous data transmission on an optical circuit on different optical carriers, and this term has become generally-accepted in the domestic literature. (The term "frequency-division multiplexing" has been retained to apply to the case of simultaneous modulation of a single optical carrier by several subcarriers.) We examined the issues of the ideology and terminology of spectral-division multiplexing in survey [4].

Research on spectral-division multiplexing in optical fibers began in the field of fiber-optics in 1979 on the impetus and under the direction of Doctor of Physics and Mathematics E.M. Dianov. A series of different studies devoted to spectral division multiplexing were the first such domestic investigations. Such research was discussed most extensively in the Candidate's dissertation of the present author [5] in 1983. Some of this research was included in survey [4] published in 1983 containing extensive and comprehensive information on spectral-division multiplexing and many concepts relating to the prospects for the future development of this field.

The present chapter reflects the primary results from research on spectral-division multiplexing carried out at the Division of Fiber-Optics between 1979 and 1983. Study [4] has pointed out that the problem of separating spectral-division multiplexed channels (transmissions at different wavelengths) is a spectroscopic problem (separation of the spectral components). Therefore demultiplexers and multiplexers often take the form of mini-spectrometers. Spectrometer operating parameters may vary significantly due to variations in size (the optical focal distances), the number of diffraction grating lines and the slit width (in our case the optical fiber ends function as the slits). In view of such freedom in selection of the parameters of the spectral-division multiplexing devices the present chapter will not include detailed information on the devices developed here, although the fundamental equipment parameters are provided for orientation.

Spectral-Division Multiplexing Links and Hardware

Mini-spectrometers. Our first study on spectral-division multiplexing resulted in the development of a laboratory multiplexed communications link in the promising 1.3 μm range [3]. Fig. 1 provides a good illustration of the primary sections of any spectral-division multiplexed link: radiation sources 2 radiating at different wavelengths; spectral multiplexer 4; optical fiber circuit 6; spectral demultiplexer (elements 7, 8, 12, 13); and photodetectors 9. The spectral-division multiplexing capabilities are most significant with monomode radiation sources (the greatest number of channels are achieved in this spectral range). Therefore 1.3 μm semiconductor lasers were activated in a monomode condition. It was demonstrated from experimentation employing a 500 m optical fiber that it is possible to multiplex channels with a 10 nm spectral interval with a crosstalk of less than -20 dB (optical power level; this figure will be used henceforth). It is possible to reduce the number of channels in this link to 10 for the case of a multimode silica glass optical fiber. The spectral channel demultiplexer in the link is an autocollimation mini-spectrometer.

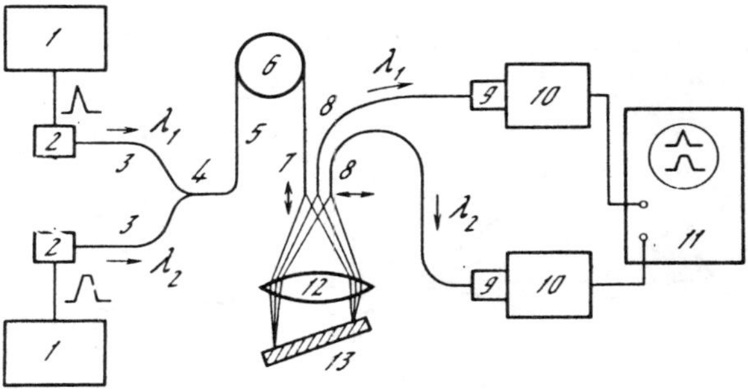

Fig. 1. Experimental fiber-optic communications link employing spectral division multiplexing in the 1.3 μm range
1 - G5-15 pulse generators; 2 - semiconductor heterolasers; 3 - injection optical fibers; 4 - fiber-optic coupler; 5 - optical fiber; 6- optical fiber bobbin; 7 - radiating optical fiber; 8 - detection optical fibers; 9 - FD9E111 photodiodes; 10 - amplifiers; 11 - S1-34 oscillograph; 12 - "Jupiter-8" objective; 13 - diffraction grating

This same mini-spectrometer (Fig. 2) was used to solve the "reverse spectroscopic" problem of integrating the luminous fluxes at different wavelengths into a single dataflow. With this multiplexing technique the radiation lines of the sources may overlap (particularly for the case of light-emitting diodes), which causes unacceptable crosstalk levels. Preliminary optical filtering of LED radiation (linewidth narrowing) was employed prior to the multiplexing process in order to avoid this phenomenon; multistage multiplexing circuits with a set of special light filters are normally used for this purpose. We employed a simple mini-spectrometer to narrow the LED lines in the multiplexing process from 43 nm (at half amplitude) to 13 nm [6] (optical fibers with a 400 μm core diameter were employed). The spectral filtering principle is clear from Fig. 2: only the central spectral components of the radiation from two different optical fibers will impinge optical fiber 1, since each of the signals falls within the spectrum of the diffraction grating. The proposed filtering method makes it possible to replace the set of special interference light filters with a single diffraction grating thereby allowing tuning through the filtering bands by simple displacement of the optical fiber ends in the focal plane of the object.

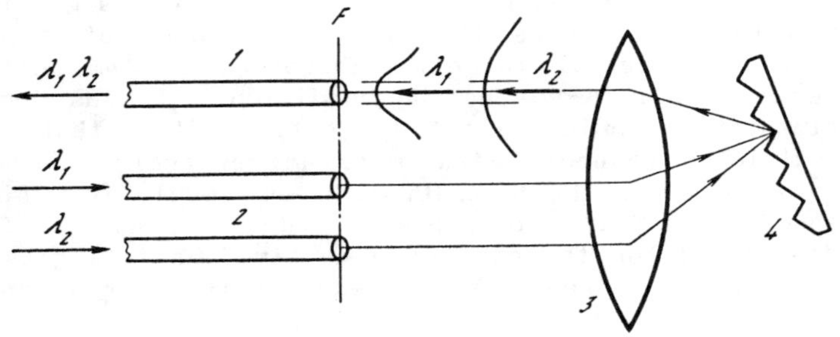

Fig. 2. Spectral-division multiplexing scheme
1 - communications link optical fiber; 2 - optical fibers providing radiation from the LEDs; 3 - objective; 4 - diffraction grating

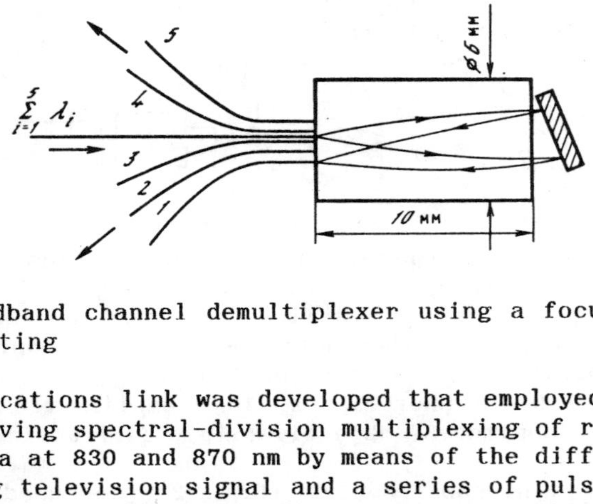

Fig. 3. A broadband channel demultiplexer using a focusing rod and a diffraction grating

A communications link was developed that employed a 100 m optical fiber achieving spectral-division multiplexing of radiation from 2 LEDs with maxima at 830 and 870 nm by means of the diffraction grating [6]. An analog television signal and a series of pulses with a wavefront duration of 0.1 μs were simultaneously transmitted at different wavelengths. The channel demultiplexer which is identical to the multiplexer performs spectral separation of the channels. Without preliminary filtering of the LED radiation it is impossible to eliminate intolerable crosstalk with a 40 nm spectral interval between the line centers. After filtering the noise level drops and this interval may be reduced to 20 nm. Here the influence of a temperature shift in the LED radiation lines is virtually eliminated, since the filtering band is fixed by the multiplexer design. By narrowing the spectral

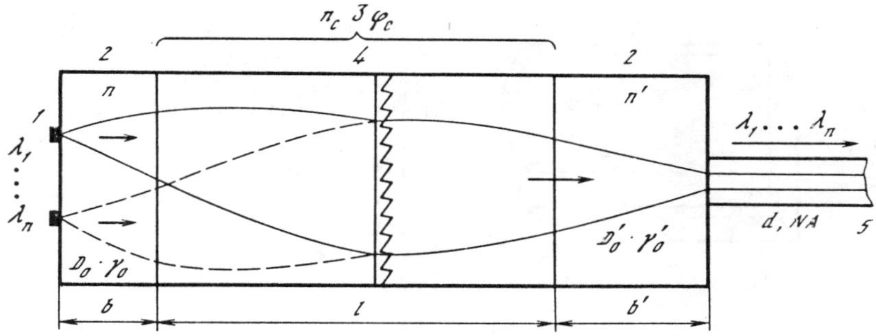

Fig. 4. Design of a spectral parallel multiplexer for monomode semiconductor lasers for feeding a monomode optical fiber based on a matching focusing rod with a phase diffraction grating
1 - laser; 2 - homogeneous rods; 3 - focusing rod; 4 - phase diffraction grating; 5 - monomode optical fiber

band of each channel it was possible to achieve in our system spectral-division multiplexing of radiation from 5 LEDs in the spectral window of the optical fiber near 0.8 μm. This increases the total bandwidth of a single optical fiber with a silica glass core and polymer coating from 25 to 125 MHz·km, thereby allowing simultaneous transmission of 5 television signals on a single step refractive index optical fiber over a 4 km distance.

Microoptic devices. Microoptic elements make it possible to fabricate demultiplexers/multiplexers as compact, single-unit devices. We investigated a broadband demultiplexer for spectrally-multiplexed channels based on a parabolic index focusing rod [1] (Fig. 3). A short focal distance was used for the rod (~5 mm) allowing significant linear back dispersion of the device. As a result it was possible to cover a broad spectral range (from 0.6–1.1 μm) with efficient concentration of the energy of the broad lines of the optical carriers (LED radiation) at the ends of the detection optical fibers.

Launching even a single optical carrier into a monomode optical fiber is a complex process due to the small core of the fiber (~ 5 μm) and a small aperture (~0.06); the efficient launching of several carriers is even more difficult. We proposed a single-unit microoptic channel multiplexer employing a phase diffraction grating within the rod [5] (Fig. 4), allowing parallel injection of several carriers into a monomode optical fiber with simultaneous matching of the output of several semiconductor lasers to the input of the monomode optical fiber. The rod parameters for such a matching process may be rather easily calculated using the formulae obtained from a geometric optics approach to investigating laser beam transformation in the focusing rod [7].

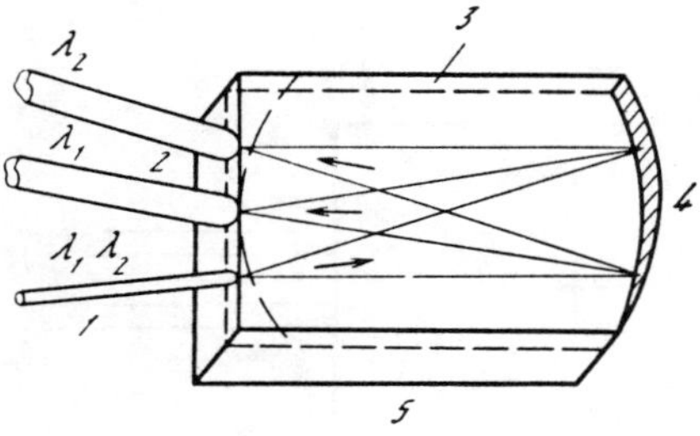

Fig. 5. A plane fiber-optic demultiplexer employing a concave diffraction grating
1 - input optical fiber; 2 - output optical fibers; 3 - plane optical fiber; 4 - end etched aluminum-coated concave grating; 5 - silica plate. The hatched arc represents part of the Rowland circle

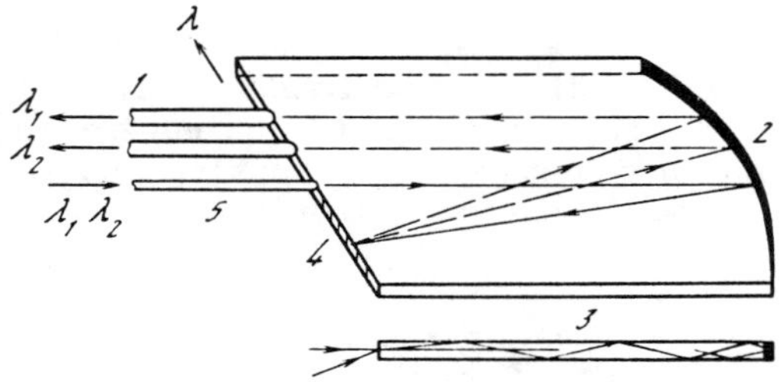

Fig. 6. Spectral division multiplexed channel demultiplexer employing a light waveguiding plate
1 - output optical fibers; 2 - rounded silver-coated end; 3 - silica glass plate; 4 - diffraction grating; 5 - input optical fiber

Integrated optical devices. The future of optical terminals for fiber-optic links is closely related to integrated optics. As a result significant attention has been devoted to investigations of integrated spectral-division multiplexing devices.

A channel demultiplexer based on a plane diffusion multimode optical fiber with a concave diffraction grating has been fabricated and tested [8] (Fig. 5). Initially the concave grating was generated directly on the concave optical fiber cut by means of laser beam interference. The demultiplexer is a plane waveguide integrated analog of a three-dimensional spectrometer employing a concave grating. The demultiplexer may separate optical carriers with a 43 nm spectral range and a 15 dB optical power crosstalk (corresponding to a 30 dB electrical power level). The spectral characteristics of the demultiplexer may be enhanced by approximating the profile of the optical fiber front cut to a Rowland circle. An investigation of the aperture characteristics of the monomode and multimode optical fibers employed in the demultiplexer and the radiation from the zero diffraction maximum at the cut of a 70 μm optical fiber has revealed that a significant source of losses is the mismatch in the numerical apertures of the optical fiber and the waveguide in the plane perpendicular to the waveguide. Moreover, the distribution of the radiation aperture at the fiber end is symmetrical with respect to its axis, while the receive aperture of the fiber drops across the end cut from the glass/air interface to the boundary of the diffusion region.

We overcame this difficulty by converting from a graded-index diffusion optical fiber to an optical waveguiding plate with a high numerical aperture that is constant across the plate cut. An optical waveguiding plate-based channel demultiplexer was proposed and implemented [9-11] (Fig. 6). A fused silica plate hundreds of microns thick was directly used as a plane waveguide in the demultiplexer. The reflecting coating on the rounded cut and the plane grating on the flat cut (coinciding with the cut/mirror focal plane) makes the plate a quasi two-dimensional analog of a three-dimensional autocollimation spectrometer employing a spherical mirror and a plane grating. The device separates six carriers in the 0.8-1.1 μm range with single channel losses of less than 3 dB [10] (Fig. 7).

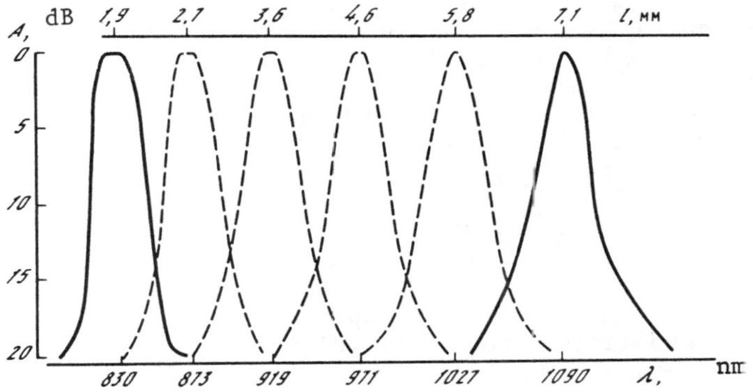

Fig. 7. The spectral power distribution in the six output optical fibers from the plate channel demultiplexer. A - attenuation from the maxima
l - distance between the optical fiber axes and the plate axis. The monochromator slit was used as the source of tunable radiation with a ~2 nm linewidth at 0.5 amplitude.

The use of the wafer as a waveguide has a number of advantages. First the problem of matching the apertures of the optical fiber and the plane waveguide is solved. Virtual total injection of luminous energy to the demultiplexer from all types of optical fibers is achieved: from monomode fibers to step refractive index fibers with a core diameter of hundreds of microns. Second since the size of the diffraction images (or any other images) is limited in one of the dimensions of the plate, the three-dimensional problem of reducing aberrations (in regular optical systems) is in our case reduced to the significantly simpler two-dimensional problem. If the regular plane schematic images of three-dimensional optical devices are simply a two-dimensional idealization of the path of rays in three-dimensional space, in our case this idealization virtually is identical to reality (ignoring the zig-zag propagation of rays between the wide plate faces). The third advantage (and probably the most important advantage) is that it is possible to fabricate different three-dimensional optical devices in a two-dimensional integrated design.

Monomode distributed couplers. An interesting physical problem is the redistribution of energy between monomode optical fibers due to their distributed coupling. Study [112] was the first to experimentally implement a design in which the cores of bent monomode optical fibers with fixed radii of curvature were similar. Such a design not only allows controlled side polishing of the optical fibers (down to the core), but also makes it possible to formulate a mathematical model of distributed coupling. Study [5] has investigated such coupling from the viewpoint of its application in spectral-division multiplexing devices. It follows from the wavelength dependence of energy redistribution among the optical fibers that if radiation at one wavelength is injected to one of the optical fibers and radiation with a wavelength approximately 10 nm higher is injected to the other optical fiber, the first signal will jump to the second optical fiber in a single step, while the second signal will end up at the output of the second optical fiber after a round trip. Thus optical carriers may be multiplexed in this coupler structure in a monomode optical fiber (they may be demultiplexed in the return path of the rays).

Study [12] was the first to experimentally achieve very low loss (~1 dB) energy transfer from one monomode optical fiber to another. 50% energy transfer is also possible (by dividing the luminous flux in half between the fibers). As a result a spectral distributed coupler may also be used in interferential multiplexing (demultiplexing) systems for optical carriers separated by a few nanometers.

Duplex Spectral-Division Multiplexed Communications Links

Spectral-division multiplexing is most suitable from the engineering and economic standpoint for long-distance communications links, since laying additional fiber in such systems significantly increases the overall system cost while the spectral width of the optical fiber per carrier drops with distance. Obviously there is a demand for duplex communications on a single optical fiber. Therefore we have developed and investigated two experimental duplex spectral-division multiplexed links. In both systems the optical fibers were 5 km long. In laboratory conditions such a length was achieved by welding together kilometer-long segments of multimode silica glass step refractive index optical fibers.

We first considered a nonspectral-multiplexed duplex link (using identical carriers) [13]. Experiment revealed that such a link has significant crosstalk interference and poor reliability, since the legitimate signal traveling along the path in each channel is comparable to the parasitic signal from the other channel (due to connectors allowing signal reflection and due to Rayleigh backscattering).

Spectral division multiplexing makes it possible to reliably suppress crosstalk by means of spectrally-selected elements.

Duplex communications link employing semiconductors with different bandgap. A reliable duplex communications link using a single optical fiber may be achieved by semiconductors with different bandgap for use both as photodetectors and as a light filter. Two optical transmission bands in our optical fibers (at 0.8 and 1.3 μm) were helpful.

The physical principle of our duplex communications system is illustrated by Fig. 8 [14] which shows the ratio of the bandgap of germanium (employed in the photodetector at 1.3 μm) and silicon (used in the photodetector at 0.83 μm and as a light filter absorbing this radiation before the germanium photodetector). Fig. 8 also shows the bandgap ratio and the corresponding photon energies, revealing the "blindness" of the silicon detector to radiation at 1.3 μm. Each end of the link has a fiber-optic splitter with a photodetector and a radiator [5] (Fig. 9). The mutual isolation of the channels is clear from Figs. 8 and 9.

Our first duplex communications link was based on a reliable physical principle: a sharp differential in the spectral characteristics of the semiconductor materials, and therefore it was extremely reliable and provided complete channel isolation without requiring special light filters. The diagram in Fig. 8 may be treated as a general configuration: it is possible to construct a spectral-division multiplexed link by the physical principle shown in the diagram using any pair of semiconductors with different bandgaps (for example, for duplex communications at 1.3 and 1.55 μm).

Fig. 8. The ratio of bandgaps of germanium and silicon and the photoionization energies used in duplex communications

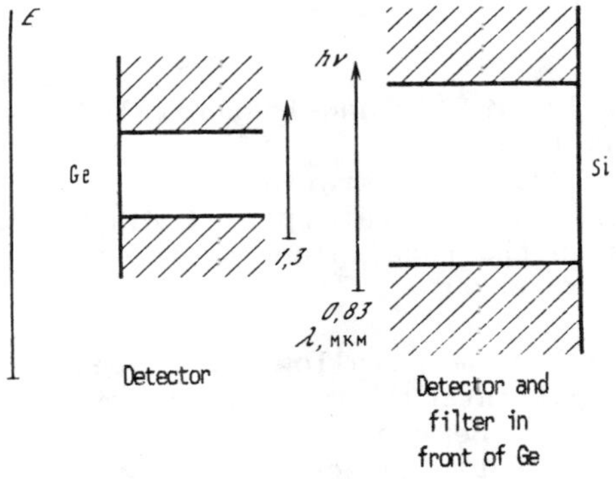

A multichannel mini-spectrometer duplex communications link. We succeeded for the first time in implementing a multichannel spectral division multiplexed communications link at 1.3 μm employing dispersive elements [13] (Fig. 10). Its principle of operation involves a configuration where a single diffraction grating at each end of the link isolates the two optical carriers propagating along the 5 km optical fiber and simultaneously launches two other carriers into the

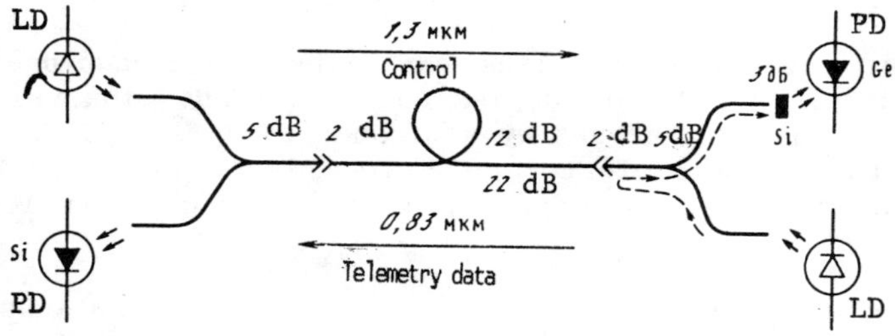

A duplex spectral-division multiplexed 5 km link
Channel isolation is achieved by using photodiodes with different spectral sensitivities and a single light filter; *ЛД* - laser diode; *ДП* - photodiode

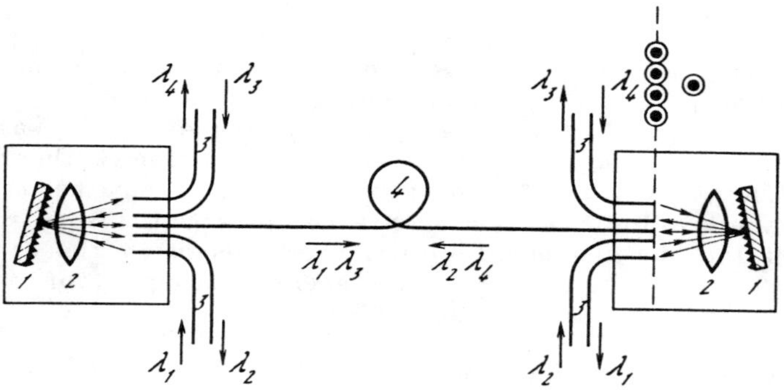

Fig. 10. Multichannel spectral-division multiplexed duplex communications link
1 - diffraction grating; 2 - objective; 3 - input and output optical fibers; 4 - linear optical fiber. The position of the optical fiber ends in the focal plane of the objective is shown on the upper right

optical fiber, followed by spectral filtering of the latter. The spectral interval between the co-moving carriers was 17 nm. Ten carriers may be used in each direction with all 20 optical carriers occupying a 100 nm spectral range. Replacing the step refractive index optical fiber with a core diameter of 50 μm by a graded-index optical fiber with an identical core diameter in no way alters the operation of the optical terminals, while the bandwidth in each channel may be significantly expanded.

Conclusion

A series of optical investigations of spectral division multiplexing in optical fibers was carried out. Channel multiplexers/demultiplexers based on mini-spectrometers, microoptics (a focusing rod), integrated optical devices (a plane waveguide, and optical waveguiding glass plate), distributed coupling between monomode optical fibers, and based on a pair of semiconductors with a different bandgap were proposed and implemented. The applicability of these devices was tested in experimental spectral-multiplexed fiber-optic communications links, specifically a multichannel duplex 5 km link operating in the 1.3 μm range.

In sum the derived results make it possible to replace RF channel multiplexing by spectral-division multiplexing using the simplest optical fibers with very high data traffic and to reduce their number in the transmission circuit.

Bibliography

1. Belovolov, M.I., Dianov, Ye.M., Kosyakov, V.I., Kuznetsov, A.A. "A broadband channel demultiplexer for spectral-division multiplexed fiber-optic communications links" IZV. AN SSSR. SER. FIZ., 1980, V. 44, p. 2075-2078.

2. Tomlinson, W.J. "Wavelength multiplexing in multimode optical fibers" APPL. OPT., 1977, V. 16, p. 2180.

3. Belovolov, M.I., Gorelenok, A.T., Dianov, Ye.M., et al. "A prototype of a spectral-division multiplexed fiber-optic communications link in the 1.3 μm range" KVANTOVAYA ELEKTRON., 1979, V. 6, p. 2487-2490.

4. Dianov, Ye.M., Kuznetsov, A.A. "Spectral division channel multiplexing in fiber-optic communications links: (a survey)" Ibid., 1983, V. 10, p. 245-264.

5. Kuznetsov, A.A. "Spectral-division multiplexing techniques in optical fibers" Dissertation for the Candidates Degree of Physics and Mathematics 01.04.04, Moscow, 1983, 169 p.

6. Belovolov, M.I., Dianov. Ye.M., Kuznetsov, A.A., Kuznetsov, A.V. "An experimental fiber-optic communications link employing spectral-division multiplexing of LED emission by a diffraction grating" KVANTOVAYA ELEKTRON., 1982, V. 9, p. 418-420.

7. Kuznetsov, A.A., Tsubulya, A.B. "Calculating the parameters of a laser beam transmitted through a focusing rod" Ibid., 1983, V. 10, p. 430-432.

8. Dianov, Ye.M., Juznetsov, A.A., Sychugov, V.A., Tulatskova, T.V. "A demultiplexer for spectral-division multiplexed channels using a plane multimode waveguide" Ibid., 1981, V. 8, p. 384-386.

9. Belovolov, M.I., Dianov, E.M., Kuznetsov, A.A., et al., "Optical demultiplexer with a glass slide" Third Intern. Conf. Integrat. Optics and Opt. Fiber Commun., 1981, 27-29 Apr. Digest Techn. Pap. OSA/IEEE. San Francisco (USA), 1981, p. 66-67.

10. Belovolov, M.I., Dianov, E.M., Kuznetsov, A.A., et al. "A spectral-division multiplex channel demultiplexer for fiber-optic communications links" PIS'MA I ZhTF, 1981, V. 7, p. 1508-1512.

11. Belovolov, M.I., Dianov, Ye.M., Kuznetsov, A.A., et al. "A spectral-division multiplexed channel demultiplexer employing a light waveguiding plate" KVANTOVAYA ELEKTRON., 1982, V. 9, p. 429-432.

12. Parriaux, O., Gidon, S., Kuznetsov, A.A. "Distributed coupling on polished single-mode optical fibers" APPL. OPT., 1981, V. 20, p. 2420-2423.

13. Alferov, Zh.I., Belovolov, M.I., Gur'yanov, A.N., et al. "A multi-channel duplex fiber-optic communications link at 1.3 μm" KVANTOVAYA ELEKTRON., 1982, V. 9, p. 1698-1700.

14. Kuznetsov, A.A. "Spectral division multiplexing techniques in optical fibers": Dissertation for Candidates Degree of Physics and Mathematics, 01.04.04, Moscow, 1983, 26 p.

Ultrashort Pulses in Fiber Stimulated Light Scattering Lasers

Ye.M. Dianov, A.N. Pilipetskiy, A.M. Prokhorov, V.N. Serkin

Abstract: The problem of controlling CW laser radiation by means of fiber Raman and Brillouin lasers is examined. The possibility of generating giant stimulated Brillouin pulses with a duration an order of magnitude shorter than the hypersonic relaxation time is demonstrated. The possibility for cascade generation of stimulated Raman pulses initiated by giant stimulated Brillouin pulses is investigated. The fundamental possibility for developing a fiber Raman soliton-laser is demonstrated.

The use of high-quality optical fibers in quantum electronics led to the development and fabrication of new laser types: fiber stimulated Brillouin lasers (SBM) and fiber stimulated Raman lasers (SR).

The present study discusses the primary physical mechanisms behind the formation of ultrashort light pulses in fiber SBM and SR lasers. Primary attention is devoted to an analysis of the current principal unresolved problems.

First this includes the problem of controlling the parameters of CW laser radiation. As demonstrated in this study fiber lasers may implement nonlinear conversion of monochromatic radiation (pump radiation) into a periodic sequence of powerful, short Stokes backward scattering pulses (both SBM and SR). In this case the giant SBM pulses accumulate virtually all CW pump radiation energy stored during the cavity period, and have a duration an order of magnitude shorter than the relaxation time of the hyperacoustic wave. Further possibilities for conversion of laser radiation in a "cascade SBM + SR" fiber laser are discussed; in this laser the SBM pulses generated in the cavity induce backward Raman scattering pulses.

The second problem involves controlling the characteristics of multimode lasers producing pulses in the picosecond and subpicosecond range. Indeed, the broad SR lines in glasses make it possible to generate such pulses in principle. However in our opinion the implementation of this possibility has not been given sufficient attention. Above all there has been no theory of ultrashort light pulse generation that accounts for the self-action effects in the SR distributed active medium (the optical fiber). The purpose of the present study is to fill this gap. Specifically, as demonstrated in the study, there exist at present all experimental prerequisites for implementing a fiber Raman soliton-laser.

1. Nonlinear Cascade Conversion of Monochromatic Radiation in a Fiber Raman Laser

Fiber Raman lasers may be used for nonlinear conversion of CW laser radiation into a sequence of powerful, short light pulses. The process of such radiation conversion is as follows.

We will assume that the fiber laser cavity already contains a short seed Stokes SBM pulse whose duration is less than the transit time of light through the cavity. CW pump radiation is taken from an external source and injected to the optical fiber which is the SR distributed active medium. As the feed Stokes SBM pulse propagates through the optical fiber in high gain conditions ($gI_L\ l \approx 10$ or more) conditions may be created where this pulse "accumulates" all pump energy stored in the cavity period and becomes significantly shorter in duration. We will further assume that the laser cavity mirrors reflect radiation at the Raman lasing frequencies as well. Then the SBM "giant pulse" formed in the cavity generates companion SR pulses also counterpropagating against the pump radiation and "acquiring" energy from it. In this case the giant SBM pulse "triggers" lasing in the fiber Raman laser. The Stokes SR pulses are further compressed and self-steepened from backward gain in the CW pump field. Therefore by implementing the "SBM + SR" cascade lasing state in the fiber laser it is possible to achieve nonlinear conversion of CW radiation into pulsed radiation. In this case the broad Raman gain lines in optical fibers [1] make it possible to shape and amplify picosecond and subpicosecond pulses.

We should emphasize that study [2] made the first experimental attempt in nonlinear conversion of CW radiation into pulse radiation by means of an optical fiber of sufficient length in an optical cavity. The authors of study [2] have noted that the lasing from the fiber Brillouin laser was unstable and therefore they employed an acoustooptic modulator to stabilize the sequence of short pulses.

We will consider the following model of a ring fiber laser. Assume the optical fiber is in a ring optical cavity formed by mirrors R_1, R_2, and R_3. The laser is pumped by CW radiation injected to the optical fiber through one of the cavity mirrors.

We will limit our examination to an analysis of the most interesting case where only two types of waves are excited in the cavity: the first Stokes SMB component and the first Stokes SR component. We will describe the interaction of the radiation field with the intrinsic oscillations of the medium: hypersound and the molecular oscillations - within the framework of the following system of equations for the complex amplitudes of the pump waves $E_L(z, t)$ of the first Stokes SMB components - $E_{SMB}(z, t)$ and the SR components $E_{SR}(z, t)$:

$$\frac{\partial E_L}{\partial z} + \frac{1}{v_L} \frac{\partial E_L}{\partial t} = -g_{MB}^{eff} \rho E_{SMB} - g_R^{eff} |E_{SR}|^2 E_L , \quad (1)$$

$$\frac{\partial E_{SMB}}{\partial z} - \frac{1}{v_{SMB}} \frac{\partial E_{SMB}}{\partial t} = - g_{MB}^{eff} \rho^* E_L + g_R^{eff} |E_{SR}|^2 E_{SMB} , \quad (2)$$

$$\frac{\partial E_{SR}}{\partial z} - \frac{1}{v_{SR}} \frac{\partial E_{SR}}{\partial t} = -g_R^{eff}(|E_L|^2 + |E_{SMB}|^2) E_{SR} , \quad (3)$$

$$\frac{1}{\Gamma} \frac{\partial \rho}{\partial t} + \rho = E_L E_{SMB}^* . \quad (4)$$

Here equation (4) describes the dynamics of the slowly varying complex amplitude of the hyperacoustic wave with a characteristic attenuation time $T_2 = 1/\Gamma$; g_{SMB}^{eff} and g_{SR}^{eff} are the effective SMB and SR gains averaged over the optical fiber cross-section [1].

We will solve equation system (1)-(4) with the following boundary and initial conditions:

$$E_L (z = 0, t) = E_{L0} , \quad (5)$$

$$E_{SMB} (z = l, t) = \sqrt{R} E_{SMB} (z = 0, t - T_3) , \quad (6)$$

$$E_{SR} (z = l, t) = \sqrt{R} E_{SR} (z = 0, t - T_3) , \quad (7)$$

$$E_{SMB} (z, t = 0) = \Phi (z/v) , E_{SR}(z, t = 0) = \Phi_0 . \quad (8)$$

Here T_3 is the delay of the radiation in the cavity outside the optical fiber; $\Phi(\tau)$ is Gaussian noise with a characteristic correlation time $\tau - T_2$.

Equation system (1)-(4) with initial and boundary conditions (5)-(8) completely determined the dynamics of SR lasing at subsequent times.

After a long pump pulse with a sharp leading edge is injected to the cavity, SMB lasing begins to develop from the spontaneous noise level. In the initial lasing period as long as $I_{SMB} \ll I_{L0}$ the spontaneous noise in the frequency band $\Delta\omega = \Delta\omega(t = 0)/(g_{MB}|E_{L0}|^2 l)^{1/2}$

will be filtered in the pass through the cavity and intensity fluctuations of the Stokes wave will be smoothed out: $\tau \mathcal{L}_l(t) = \tau \mathcal{L}_l(t=0)$ $(g_{MB}|E_{LO}|^2 l)^{1/2}$. As the SMB radiation intensity grows saturation ensues $I_{SMB} \simeq I_{LO}$, and a decreasing portion of pump radiation will enter the optical fiber. As a result the effective gain per pass becomes time-dependent: it is at a peak at the leading edge of the generated SBM pulse and is at a minimum level at its trailing edge. The preferential gain of the leading pulse edge causes the formation of a separate Stokes SMB pulse in saturation during the cavity period. We emphasize that this pulse separation effect at the Stokes frequency essentially follows the same physical mechanism as the amplitude modulation of pump radiation and of the reflected wave with a period equal to twice the pass period through the medium [3]. Fig. 1 shows a computer-generated detailed picture of the Stokes pulse formation process in the cavity of a fiber Brillouin laser for the case of lasing from spontaneous noise (a multimode model of Gaussian noise was employed in the calculations).

In the initial lasing stage the Stokes pulse is amplified and compressed, although with each pass through the laser cavity it leaves in its wake a powerful hypersonic wave near the front end of the optical fiber, which reflects backwards a portion of the pump radiation and therefore creates an extended trailing edge of the pulse. This residual hypersonic wave produces a situation where with each subsequent pass through the cavity a progressively lower pump intensity reaches the optical fiber and the temporal structure of the Stokes radiation is stretched out. We examined the dynamics of this process in detail in study [4]. A numerical experiment indicates that in order to convert CW radiation into pulsed radiation in fiber lasers it is necessary to "cut-off" the trailing edge of the Stokes pulse. We proposed using nonlinear bleaching of the resonant absorbing medium employed in the laser cavity for this purpose in study [4].

If the intensity of the giant SBM pulses generated in the fiber laser cavity reaches the Raman lasing threshold, a cascade Raman pulse generation mechanism may be realized.

As indicated by numerical experiment the decisive factor determining the nature of combined generation of SMB and SR pulses in a stimulated Raman laser is the ratio of the stationary Raman and Brillouin gains

$$\theta = g_R^{eff} / g_{MB}^{eff} \; .$$

With low values of the parameter $\theta \leq 10^{-2}$ (for glasses $\theta = 10^{-2}$–10^{-3} [1]) short and intense SMB pulses are generated in the initial lasing stage which then "trigger" the Raman laser (Fig. 2). When $\theta \ll 1$ the threshold SR lasing condition is satisfied only near the peak of the SMB pulse envelope, which causes compression of the Stokes SR pulses during the development process. The SR pulses are formed on the trailing edge of the intense SMB pulse. This effect - the formation of a short SR pulse at the leading edge of the SMB pulse and its

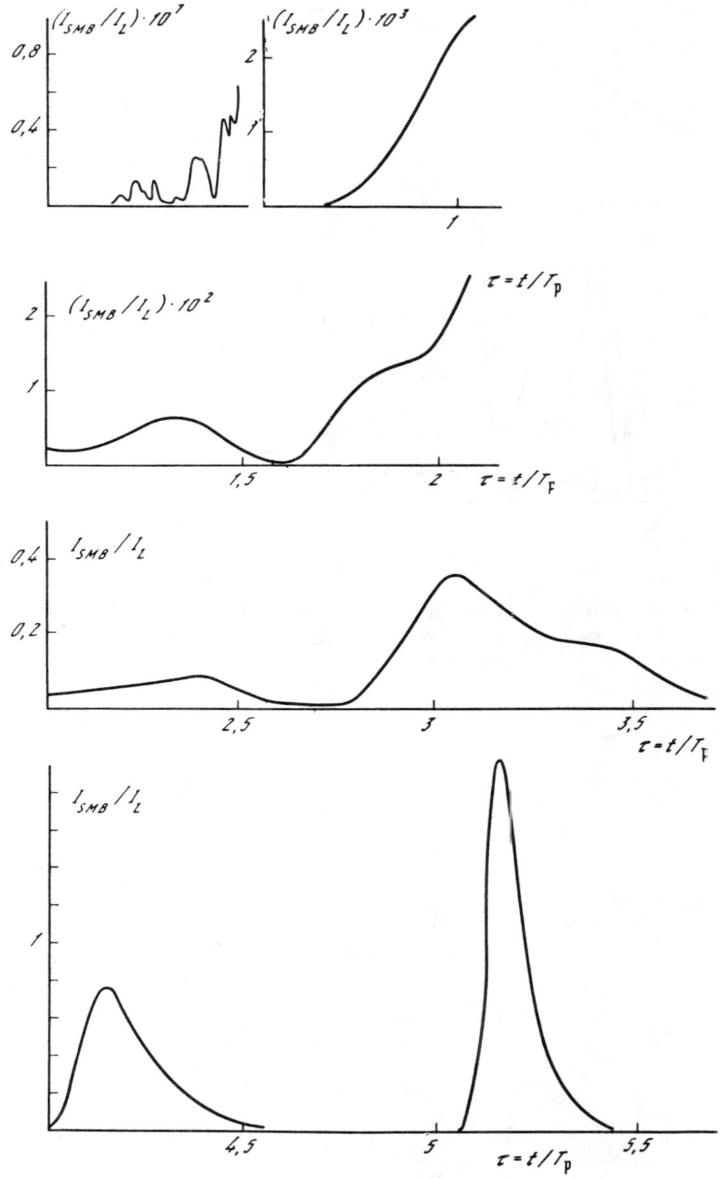

Fig. 1. The formational process of a single SMB pulse in a laser cavity for the case of lasing from spontaneous noise
Time is given in units of $\tau = t/T_p$ (T_p is the cavity period)

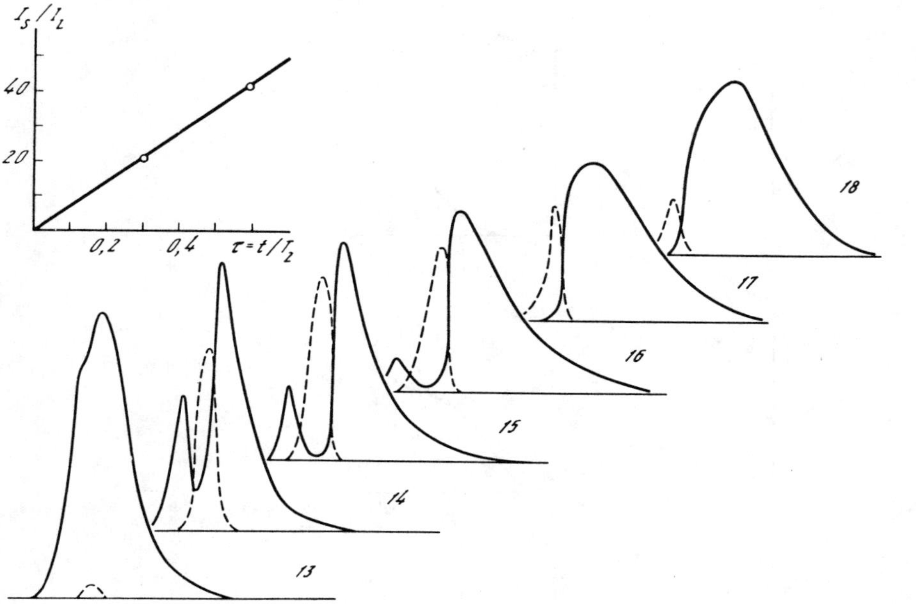

Fig. 2. The temporal structure of SMB pulses (solid curve) and SR pulses (dashed curve) after k passes through the laser cavity (numbers by the curves) for $\theta = 5\cdot10^{-3}$

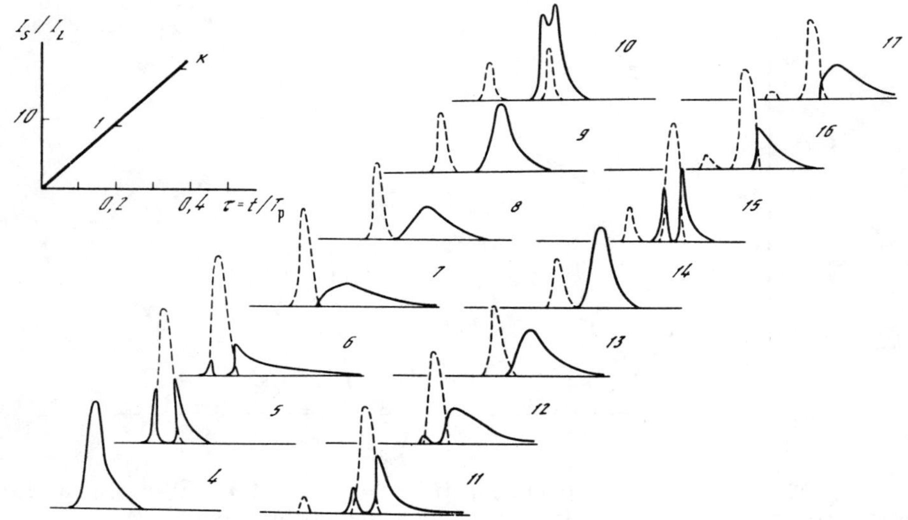

Fig. 3. The temporal structure of SMB pulses (solid line curve) and SR pulses (dashed curve) for $\theta = 10^{-1}$
The numbers by the curves indicate the value of k

subsequent "breakaway" from the pulse (Fig. 2) - is caused by the reduction in group velocity of SMB pulses due to the lag in response of the hypersonic oscillations of the medium. Moreover, the dispersion of the group velocity of pulses channeled in the optical fiber ($\upsilon_{SR} > \upsilon_{SMB}$) will also cause a "breakaway" of the Stokes SR pulse from the leading edge of the SMB pulse.

Thus the interaction of two coupled "SMB + SR" lasers when $\theta \ll 1$ produces a lasing state where the laser emits "pulsating" trains of Raman pulses (Fig. 2). As this parameter grows the SR pulses "interrupt" the generation of SMB pulses earlier than the latter can develop to their maximum values. When $\theta = 10^{-1}$ (Fig. 3) new SR pulses will arise as the SR Stokes pulse "breaks away" from the leading edge of the SMB pulse and "luminesces" at the cavity mirrors during the cavity period (Fig. 3). Due to hypersonic lag the new SMB pulse is shifted in time with respect to the preceding pulse and the laser emits trains of relatively delayed sequences of short SMB and SR pulses (Fig. 3). Competition between the SMB and SR processes results in the formation of single intense SR pulses ($\theta \sim 1$).

Thus, fiber Raman lasers have created new possibilities for controlling the parameters of CW laser radiation and make it possible to implement nonlinear conversion of monochromatic radiation into a periodic sequence of powerful, short pulses.

2. Formation of Ultrashort Pulses in Multimode Fiber Raman Lasers: Cascade Self Mode Locking, Optical Pulsons, and Solitons

We know that the lasing efficiency of a fiber Raman laser is significantly improved by placing the optical fiber and the pump laser in a common cavity [5]. This forms a complex system of two or more interacting lasers that sequentially "fire" one another. The optical fiber in this case represents the Raman distributed active medium. In the present study we will provide results from the numerical modeling of the dynamics of such a laser system in nonstationary lasing (radiation of relaxation peaks or cavity Q-switching).

At low pump intensities (below the Raman lasing threshold) the temporal structure of radiation from a multimode laser with a fiber cavity (at the lasing fundamental) takes the form of a quasiperiodic sequence of fluctuating radiation spikes generated from the interference of a large number of longitudinal modes with a random phase distribution. As the lasing intensity within the laser cavity grows it reaches a second threshold: the lasing threshold at the first Raman frequency. The development of Raman lasing in the cavity of the pump laser begins at the spontaneous scattering level which is, by virtue of its physical nature, a Gaussian process whose intensity fluctuates and it contains spikes with a characteristic correlation time determined by the spontaneous Raman linewidth. Moreover, due to the noisy nature of multimode lasing at the fundamental (pump frequency) the transition from spontaneous scattering to SR lasing is a statistical process that involves the preferential amplification of the points on the temporal intensity profile at the heterodyne frequency that transit the cavity with maximum amplitude at the pump frequency. The temporal structure of the lasing radiation from such a laser is determined by the pump level over threshold as well as the relationship between the spectral widths at the fundamental and the heterodyne frequencies [5]. The exponential nature of SR gain will therefore produce partial self-mode locking at the frequency of the first Stokes SR component. As the pump intensity grows the conditions for the lasing threshold at the subsequent heterodyne frequencies become satisfied. A cascade self mode locking effect then arises within the spectral width of each of the Stokes components causing isolation of the single SR pulses in the period of the cavity of a fiber Raman laser. Thus, the SR active medium (the optical fiber) in the cavity of a multimode laser functions as a nonlinear filter which, unlike the widely-used bleachable filters [6] discriminates among the pulses not at the fundamental frequency but rather at the heterodyne frequencies [5].

A significant feature of the physical picture of the formation of ultrashort pulses in this model is the determinant role of self-action effects of the radiation which not only significantly complicate the lasing dynamics in a fiber Raman laser, but also create the possibility for developing a fiber Raman soliton-laser.

We will examine this possibility in greater detail using a sample model of a ring multimode fiber Raman laser; this model was discussed in detail in study [5]. Unlike this study we will consider the new features in the nonlinear dynamics of a fiber laser that arise from accounting for the self-action effects. We will describe the lasing dynamics within the framework of a system of Schrodinger equations for the complex pump wave envelopes and cascade Stokes SR components:

$$2i \frac{\partial E_L}{\partial z} = -\frac{\partial^2 k_L}{\partial \omega^2} \frac{\partial^2 E_L}{\partial \tau^2} + \frac{k_L n_2(\omega_L)\alpha_{LL}}{n_0} |E_L|^2 E_L +$$

$$+ \sum_{n=1}^{N} \frac{k_L n_2(\omega_{Sn})\alpha_{L,Sn}}{n_0} |E_{Sn}|^2 E_L - ig\alpha_{L,S1} \frac{\omega_L}{\omega_{S1}} |E_{S1}|^2 E_L - i\gamma_L E_L, \tag{9}$$

$$2i\left[\frac{\partial E_{S1}}{\partial z} + \left(\frac{1}{v_{S1}} - \frac{1}{v_L}\right) \frac{\partial E_{S1}}{\partial \tau} \right] = -\frac{\partial^2 k_{S1}}{\partial \omega^2} \frac{\partial^2 E_{S1}}{\partial \tau^2} +$$

$$+ \sum_{n=1}^{N} \frac{k_{S1} n_2(\omega_{Sn})\alpha_{S1,Sn}}{n_0} |E_{Sn}|^2 E_{S1} + \frac{k_{S1} n_2(\omega_L)\alpha_{S1,L}}{n_0} |E_L|^2 E_{S1} +$$

$$+ ig\alpha_{S1,L} |E_L|^2 E_{S1} - ig\alpha_{S1,S2} |E_{S2}|^2 E_{S1} - i\gamma_{S1} E_{S1}, \tag{10}$$

$$2i\left[\frac{\partial E_{Sm}}{\partial z} + \left(\frac{1}{v_{Sm}} - \frac{1}{v_L}\right) \frac{\partial E_{Sm}}{\partial \tau} \right] = -\frac{\partial^2 k_{Sm}}{\partial \omega^2} \frac{\partial^2 E_{Sm}}{\partial \tau^2} +$$

$$+ \frac{k_{Sm} n_2(\omega_L) \alpha_{Sm, L}}{n_0} |E_L|^2 E_{Sm} + \sum_{n=1}^{N} \frac{k_{Sm} n_2(\omega_{Sn}) \alpha_{Sm, Sn}}{n_0} |E_{Sn}|^2 E_{Sm} +$$

$$+ ig \frac{\omega_{Sm}}{\omega_{S, m-1}} E_{Sm} [\alpha_{Sm; S, m-1} |E_{S, m-1}|^2 - \alpha_{Sm; S, m+1} |E_{S, m+1}|^2] - \qquad (11)$$

$$- i\gamma_{Sm} E_{Sm}.$$

Here $E_{L, Sm}$ are the complex amplitudes of the interacting waves; $\tau = t - z/\upsilon_L$ is "traveling" time with the pump emission; $\upsilon_{L, Sn}$ are the group velocities of the pulses channeled in the optical fiber; g is the gain of the first Stokes component at the peak of the amplification line; $\gamma_{L, Sn}$ are radiation line losses in the optical fiber; α_{mn} is the geometric factor from averaging the wave fields over the optical fiber cross-section [1]. The remaining conventions are standard.

The fundamental physical processes in a fiber Raman laser described by equation system (9)-(11) may be divided into active and reactive interactions. The active interactions include the cascade energy exchange between waves making it possible to compensate wave damping by the cavity mirrors and to "transport" energy from one spectral range (for example, the range of positive group velocity dispersion $\partial^2 k/\partial\omega^2 > 0$) to another spectral range (for example, to the negative group velocity dispersion range $\partial^2 k/\partial\omega^2 < 0$). Reactive wave interaction through the nonlinear addition to the refractive index $n_2|E|^2$ transposes the phase modulation of the pump pulse to the heterodyne frequency (cross-modulation) and also causes self-broadening of the pulses in the $\partial^2 k/\partial\omega^2 > 0$ range and self-steepening of the pulses for $\partial^2 k/\partial\omega^2 < 0$. The relative role of the reactive and active interactions during the lasing process in a fiber Raman laser is determined by the parameter $\theta = z_d/z_{int}$, where $z_d = G/(gI_L^{max})$ is the Raman gain length, $z_{int} = (z_d z_{nl})^{1/2}$ is the self-action length of the pulse envelopes expressed through the dispersion length $z_d = \tau_0^2/|k''_{\omega 2}|$ and the nonlinear interaction length $z_{nl} = n_0(k\alpha n_2 I_L^{max})^{-1}$. Here $I_L^{max} = I_{st}/I_{sp})$ is the mean radiation intensity at the peak of the pulse train at the fundamental lasing frequency related to the most probable value of the maximum spike in the cavity period by the relation $I_L^{max} = \langle I \rangle \ln M$, where M is the number of generated longitudinal modes.

In the initial lasing development stage $\theta \ll 1$. However, $\theta \to 1$ as pulses appear at the heterodyne frequencies (during the "cascade" self-mode locking process). This means that the model of a Raman fiber laser examined in study [5] becomes unsuitable for a correct description of lasing dynamics in such a laser system.

We will determine the range of critical laser parameters when $\theta = 1$:

$$\tau_{Sk}^{cr} = \frac{G}{g(\omega_{Sk})} \left[\frac{2\pi n_2(\omega_{Sk})k''_\omega}{\lambda_{Sk}\delta I_L^{max}} \left| \frac{\lambda_{Sk} - \lambda_0}{\lambda_L - \lambda_0} \right| \right]^{1/2}. \tag{12}$$

Here λ_0 is the "zero" group velocity dispersion point.

As indicated by (12) the critical pulse duration τ_{Sk}^{cr} is a few picoseconds. With typical laser parameters: $T = n_0 l/c = 5 \cdot 10^{-8}$ s; M = 50; $\langle I_L \rangle = 10^7$ wt/cm^2; $\eta = P/P_{th} - 1 = 10$; $\eta_0 = 1.3$ μm; $k''_\omega = 2 \cdot 10^{-28}$ s^2/cm; $g = 10^{-11}$ cm/wt; $G = 16$; $n_2 = 3.2 \cdot 10^{-16}$ cm^2/wt; $\delta = 10$ (The radiation focusing factor on the optical fiber); $\lambda_{S5} = 1.41$ μm; λ_L = 0.6 μm is the critical duration of the Stokes pulse $\tau_{S5}^{cr} = 5$ ps.

We will consider the primary results from numerical experiments devoted to the dynamics of a Raman fiber laser in the most interesting case where one of the generated Stokes SR frequencies falls within the spectral range corresponding to negative group velocity dispersion $k''_\omega < 0$.

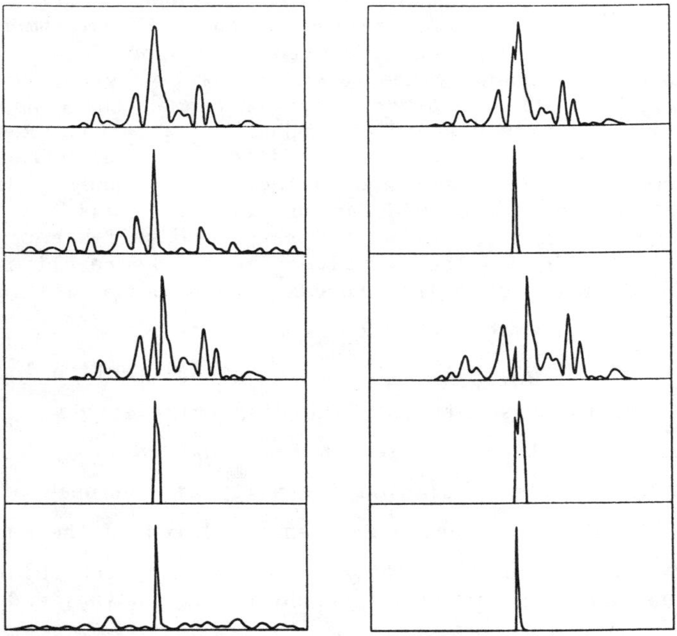

Fig. 4. "Computer oscillograms" of the cascade self-mode locking process in a fiber Raman laser at the second, third, and fourth Stokes SR components

As indicated by system (9)-(11) after the lasing threshold of the first SR Stokes component is achieved, its field serves as the "pump" for exciting the next component, etc. In the general case a group of pulses with a duration $G^{1/2}$ times less than the duration of the envelope at the preceding Stokes frequency is developed at each subsequent Stokes frequency. Pulses corresponding to the peak of the pulse envelopes at the nth Stokes frequency develop more rapidly than other pulses within each "packet" of generated pulses at the $(n + 1)$ Stokes frequency. This results in a significant drop in the duration of the pulse envelope at the cascade Stokes lasing frequencies and discrimination of certain pulse spikes in the cavity period. Fig. 4 shows the pattern of successive discrimination of fluctuation radiation spikes at the second, third, and fourth Stokes SR lasing frequencies when an optical fiber segment of sufficient length is introduced to the cavity of a neodymium glass laser ($\lambda_L = 1.06$ μm). We emphasize that the fourth Stokes SR component in the silica optical fiber falls within the "0" group velocity dispersion range $k''_\omega|_{\lambda=\lambda_4} = 0$. The next Stokes SR component ($\lambda_{S5} = 1.41$ μm) is in the spectral range corresponding to negative group velocity dispersion $k''_\omega < 0$ [7]. In this case, as predicted by numerical experiment, it is possible to implement soliton propagation of the SR pulses in the fiber Raman laser. Fig. 5 shows a typical pattern of the formation of a "pulsating" sequence of ultrashort pulses in a laser at the fifth Stokes frequency ($\lambda_{S5} = 1.41$ μm; $\lambda_0 = 1.3$ μm). The amplitude and frequency modulation of the SR pulses are responsible for the regular damping pulsations. The duration of the SR pulses (Fig. 5) amounts to a few picoseconds.

The numerical experiment makes it possible to identify the influence of the separate physical processes described by system (9)-(11) on the generation dynamics of powerful, short pulses at the Stokes frequency.

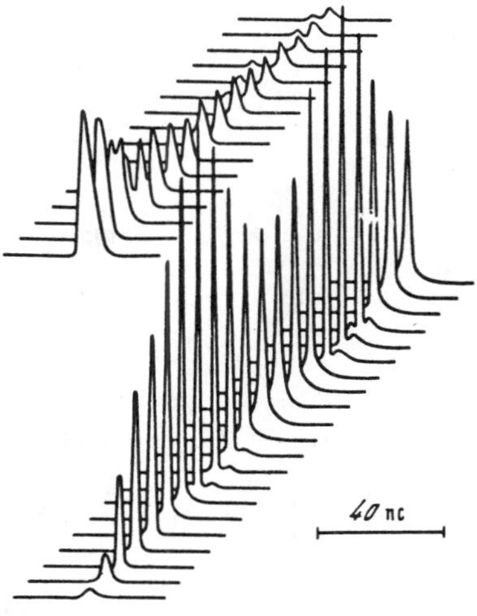

Active interaction - energy exchange between waves - is determined by the pump level over threshold η. By varying this quantity over the range $4 \leqslant \eta \leqslant 20$ in numerical experiments a quasi-stationary sequence of SR solitons was first observed which then resulted in the generation of SR pulsons with growth of η. We note that the intensity of the pulses at the fifth Stokes frequency is two orders of magnitude greater than the intensity of stationary lasing at the fundamental.

Fig. 5. The nonlinear formational dynamics of optical "pulsons" in a fiber Raman laser

In order to establish the role of reactive interaction in the development of lasing in a fiber Raman laser the calculations "neutralized" the $n_2(\omega_{Sn})$ coefficients. Such a possibility of "switching off" the reactive interaction in numerical experiments has revealed that the transposition of phase modulation of the pump emission to the Stokes frequency predicted and described in detail by V.N. Lugov in study [7] plays the fundamental role in nonlinear laser dynamics. As revealed by analysis this very effect will produce a pulsating sequence of ultrashort RS pulses (Fig. 5).

The optical scheme of a multimode fiber laser with a cavity common to both radiation at the fundamental and at the heterodyne lasing frequencies presents new possibilities for realizing various conditions for generating ultrashort discretely tunable light pulses.

In conclusion we wish to emphasize that Raman gain in optical fibers makes it possible not only to develop a Raman fiber soliton laser, but also makes it possible to amplify solitons in high-capacity fiber-optic long-range communications systems [8]. Moreover, the RS pulse transformation processes in optical fibers may be used for nonlinear conversion of solitons and for generating high-energy single-soliton pulses at the Stokes frequency [9].

Bibliography

1. Stolen, R.H. "Fiber Raman lasers" FIBER AND INTEGRAT. OPT., 1980, V. 3, No. 1, p. 21.

2. Kawasaki, O.S., Johnson, D.C., Fujii, Y., Hill, K.D. "Bandwidth-limited operation of a mode-locked Brillouin parametric oscillator" APPL. PHYS. LETT., 1978, V. 32, No. 7, p. 429.

3. Johnson, R.U., Marburger, J.H. "Relaxation oscillations in stimulated Raman and Brillouin scattering" PHYS. REV. A, 1971, V. 4, No. 3, p. 1175.

4. Dianov, Ye.M., Pilipetskiy, A.M., Prokhorov, A.M., Serkin, V.N. "Nonlinear conversion of CW radiation into a sequence of powerful and short pulses in fiber lasers" PIS'MA I ZhETF, 1985, V. 41, No. 8, p. 323.

5. Isaev, S.K., Korntsenko, L.S., Kravtsov, N.V., et al. "The lasing dynamics of a fiber Raman laser" ZhETF, 1980, V. 79, p. 1239-1256.

6. Zel'dovich, B.Ya., Kuznetsova, T.I. "Ultrashort light pulse generation by means of lasers" USPEKHI FIZ. NAUK., 1972, V. 106, p. 47.

7. Lugovoy, V.N. "Stimulated Raman radiation and frequency scanning in an optical fiber" ZhETF, 1976, V. 71, No. 4, p. 1307-1319.

8. Mollenauer, L.F., Stolen, R.H., Islam, M.N. "Experimental demonstration of soliton propagation in long fibers loss compensated by Raman gain" OPT. LETT., 1985, V. 10, No. 5, p. 229-231.

9. Vysloukh, V.A., Serkin, V.N. "Nonlinear conversion of solitons in optical fibers" IZV. AN SSSR. SER. FIZ., 1984, V. 48, No. 9, p. 1777-1781.

SUBJECT INDEX